THE
FARMALL
DYNASTY

The Story of the Engineering and Design That Created International Harvester Tractors

LEE KLANCHER

Published by:

**671
Press**

**2368 Hadley Ave. N.
St. Paul, MN 55128
www.671press.com**

Version History
671 Press edition published October 2008
Originally published by Motorbooks International as *International Harvester Photographic History*, October 1996

Klancher, Lee M. 1966-
The Farmall Dynasty/Lee Klancher

ISBN-13 978-0-9821733-0-5
ISBN-10 0-9821733-0-X

Design and layout by Cory Anderson
671 Press logo design by Micah Edel

On the Cover: The Raymond Loewy-designed IH logo on a 1954 Super M-TA.

On the Back Cover: The last of the tractors built by IH were the 50 and 30 series.

Contents

This book is dedicated to my late grandfather,

Paul C. Klancher,

who told me to do what I love.

Introduction

When the International Harvester Company (IHC) formed in 1902, some of the best and brightest engineers of the time were brought together, eventually under one roof. That group quickly took a leadership role in the rapidly forming tractor industry, an IHC tradition that continued until International tractors were no more.

The early IHC tractors were remarkable in both their quality and sheer crudity of construction. Using total-loss lubrication and hit-and-miss ignitions, the early Titans and Moguls harnessed internal combustion in the form of a heavy, difficult-to-operate, and low-horsepower machine. The fact that they worked at all is just as amazing as the thought that farmers could actually get anything done with these crude pieces of equipment.

The first machine to showcase the talents of the IHC engineers was the International 8-16, a machine that was truly ahead of its time. It was the first mass-produced tractor to be equipped with a power take-off, and only supply and manufacture difficulties kept it from being a runaway success.

The next machines of note were the McCormick-Deerings, the 10-20 and 15-30: designed by committee, ruggedly constructed, well-built, and economical to produce in quantity. They were leading sellers in their time, and carried the ball well against the Fordson until the revolutionary Farmall appeared a few years later.

The Farmall may have been IHC's crown jewel. It changed the way farmers viewed the tractor, and became the standard tractor design for the following 30 years. The industry had known for some time that the tractor that replaced the horse and could do it all—turn a belt, pull a plow, and cultivate—would be a big winner. The Farmall was that and more. Its gangly looks and questionable pedigree made it a hard sell to management, but its performance won over farmers and—eventually—International executives.

The next great leap was the Letter Series tractors, which incorporated lots of small improvements that added up to a timelessly graceful and useful line of tractors. The ultimate compliment to these machines is the multitude of battered, mud- and manure-splattered Letter Series machines still working the farm today. Very few 50-year-old machines have worked as hard and long as Letter Series tractors.

After the Letter Series, the company continued its tradition of innovative products, although the great strokes had been painted. The tractor was thereafter refined rather than revolutionized, with advances like shifting on the fly, independent power take-off, more powerful and versatile hydraulic systems, more horsepower, and four-wheel-drive finding their way to the farm.

Flashes of brilliance kept coming, though. The Cub and Cub Cadet put IHC tractors onto the home and hobby farm. The Model 706 and 806 tractors of the 1960s were high-quality examples of how good a high-horsepower tractor could be, and the 2+2 four-wheel-drive tractors brought maneuverability and unified production to articulated machines.

Tragically, high technology wasn't enough to combat changing times. Tractors were so good that farmers didn't need to replace them every 2 or 5 or even 10 years. When hard times hit, the 10- or 20-year-old tractor was overhauled rather than replaced, and manufacturers were crushed by a sagging market.

The times forced every manufacturer but one into mergers. John Deere, not IHC, was the only company left standing when the smoke cleared. In October of 1984, the announcement was made that IHC's agricultural division was to be merged with the Case tractor division to form Case-IH.

The tradition of the International Harvester Company tractors lives on, however, in the hearts and minds of thousands of former employees who made IHC their lives and with millions of farmers who benefited from the company ingenuity. Thanks to restorers and enthusiasts across the globe, today you can still see the tractors as they appeared in showrooms and fields more than 50 years ago.

Prologue

When Titans Collide

The McCormicks and Deerings Join to Dominate

*"Power farming really began to appear in 1831
with the invention of the reaper.
Its story is perhaps not yet finished, even with the twentieth century
and the invention of the tractor."*
—Cyrus McCormick III, from Century of the Reaper

The story of the International Harvester Company most often begins with Cyrus McCormick I, the man who started the McCormick Reaping Company, and the man who is credited with inventing the reaper. While this hangs the story on a very capable pair of shoulders, it is a simplified version. For one, it is unclear who actually invented the reaper. The closer one looks, the more blurry the lines become.

Despite the best efforts of Cyrus and his grandson, Cyrus McCormick III's to tell the world that Cyrus I invented the reaper, it may have been Obed Hussey who built the first workable reaper. Although Patrick Bell was said to have been using such a device in 1828. It will probably never be completely clear who invented the reaper. Too many people with vested interests loudly proclaimed that they were the genius behind the reaper. Even if one person did invent the reaper, that person was helped along by previous lessons. Inventions are always collaborations of a sort, with the person who first puts the pieces together and becomes famous for it being tagged as a genius. The invention of the reaper is an especially tawdry tale, with several men loudly proclaiming in books, magazines, and on street corners that they invented the reaper and, by the way, have a model for sale.

The story begins with McCormick's father, Robert, who experimented with a grain harvester as early as 1816. Cyrus would have been 7 years old when his father field-tested the machine. By age 21, young Cyrus had built and patented a self-sharpening plow. The story goes that young Cyrus took charge of the reaper's development in 1831 and made the machine workable. He was said to have demonstrated a working model of the reaper in the same year, 1831, but he did not file a patent at that time. Certainly, Cyrus was a bright young star with the smarts and determination to pull such a feat off.

McCormick continued to improve upon the machine until 1834, when McCormick found a photo of Hussey's reaper patent in Mechanic's Machine magazine. When he saw that there was competition in the field, McCormick immediately filed a patent on June 21, 1834.

McCormick would show that he was not one to back out of a challenge, a characteristic that would endure in the family and become a defining part of the International Harvester culture. McCormick wrote a letter to *Mechanic's Machine* challenging Hussey to a public competition between the two reapers. Hussey did not have a machine available, and the challenge went unanswered. In 1843, however, the two met on the field to settle the business, reaper-to-reaper. The McCormick machine was declared the better of the two because of its superior performance in wet grain. Thus the Reaper Wars had begun.

Early on, the war was fought mainly in court. McCormick would sue anyone who produced a machine anything like his own. He didn't always win, but he did cause the buyer to consider the fact that if the offending manufacturer lost, the farmer's piece of equipment could be dispossessed by the court.

By the 1850s, reaper sales began to climb enough to attract more and more competitors, and McCormick took his battles back to the fields. These "competitions" truly characterized the savagery of the day, as competitors sabotaged each other's machines, cussed each other roundly on

The first public test of Cyrus McCormick's reaper is said to have taken place in July 1831. McCormick didn't actually market it until the 1840s.

the fields, and occasionally dropped all pretense and broke into a good old-fashioned brawl. None of the machinery was especially reliable, so failures were common, leaving the fields covered with broken reapers and foul-tempered (and often bloodied) salesmen.

J.W. Wisehart worked as a salesman for Osborne, Deering, and others before joining the International Harvester sales staff in 1902. He recalled the field trials as all show, and no trial:

"During the early years of my harvester experience, the field trial was an important part of the sales argument to the farmer. We never entered a trial unless we had it fixed and won beforehand. That was the idea of them all."

The farmers of the day accepted this mayhem as standard business, and bought reapers like crazy. Of course, the farmers didn't pay cash.

For the most part, they couldn't. Farmers of the day made their money when their crops came in, so machinery (and most everything else) was bought on credit. They relied on what was basically one paycheck a year. If the weather was bad (and any good farmer will tell you it's always bad), the paycheck was less than hoped. If the weather was truly awful, there was no paycheck. And if the farmer didn't get paid, the reaper salesman was left holding the bag.

Getting farmers to pay their notes on machinery was a big issue with the companies of the day. Selling machinery in a buyer's market was only half the battle; extracting the money from poor and famously independent farmers was just as important. One company representative recalled the difficulty of tracking down salesmen who didn't collect on their sales:

"One interesting phase of my work was the straightening out of accounts of general agents who were behind with their collections and lax in their business methods . . .. This work was difficult and at times dangerous because agents behind in their accounts were frequently hostile to my activities and many of them refused to cooperate at all. I remember on one occasion the agent happened to be a U.S. Marshal. I called at his office and told him that I wanted to check up his collection lists with him and take with me his cash received by him from the McCormick notes. When he refused to discuss the matter and indicated he was not prepared to make a settlement I told him that I had an officer downstairs ready to serve a warrant for his arrest if he tried to depart before reaching an understanding with me. However he left the room anyway and waved aside the man who was supposed to stop him. The latter, one of his deputies, was afraid to serve the warrant. I followed the agent across the street and into a building. Finally he drew a gun, which he carried in his capacity as a Marshal, and stuck it against my ribs and I stepped aside and away he went. We finally located him in Canada."

Two of the companies that prospered in these hard times were the McCormick Harvesting Machinery Company and the Deering Harvester Company. Deering fared better at getting farmers to pay their notes, but both companies were highly successful and were fierce competitors.

The Deering company was managed by the Deering family, Charles, William, and James. They were fair and reasonable men who ran a tight ship. One of the driving forces behind the Deering company was John F. Steward, a controversial man who was something of a thorn in the McCormicks' side. Steward was heavily involved in the grain harvesting industry, and became the major detractor to the claim that Cyrus McCormick invented the reaper.

A Deering employee described Steward as follows: "In appearance and manner he was uncouth and very difficult to get along with. He seemed to make an enemy of almost everyone with whom he came in contact."

Rodney B. Swift, a long-time McCormick employee, vividly described the times: "In the period between 1879 and 1902, much rivalry existed between McCormick and other companies. Competition was particularly keen with the Deering Company. The latter vigorously pushed its machines into every place grain and grass was harvested and fought for sales, challenging the other makers to trials and took every opportunity to attack the McCormick machines in its catalogues, circulars, and other advertising matter. John F. Steward, who was a prolific inventor of trivial devices and who was head of the Patent and Experimental Division at the Deering Company, wrote and published numerous articles in the Deering Farm Paper and friendly trade papers. Steward had induced the Deerings to bring suit against the McCormick Company in the U.S. courts on some of his inventions and we defeated them. In his zeal to belittle McCormick, he gave the honor to Hussey, Bell, Ogle, and anyone else for whom he could put up a shadow of an argument."

The competition between Deering and McCormick was fierce at the turn of the century, but greater still was their shared fear of a changing market. The leaders of both companies recognized that they were at risk as independent entities, while as one dominant company they could survive and prosper.

According to Cyrus McCormick, Jr., in *The Century of the Reaper*, the Deerings and the McCormicks met several times to discuss the possibility of merging, but could not come to terms.

From the glimpse at the nature of these men afforded by the paper trails they left, it is clear that the leaders of these companies were confident people accustomed to tremendous power and to having things go their way. Packing such colossal egos around one bargaining table must surely have made sparks fly, and one doubts the gentlemanly accounts of these early meetings found in Cyrus McCormick III's book, *Century of the Reaper*. It is impossible to discern what kept these two powerful companies from completing the merger themselves, but it is easy to speculate that the meetings were wrestling matches between some of the most powerful, stubborn, and shrewd business leaders of the day.

Yet the prize of ruling the market—even if this power were shared with former competitors—was too great even for the proud, independent people that drove International Harvester to pass up. The combination of a slowing market and the allure of controlling the industry resulted in consummation of "the deal" and the creation of the International Harvester Company. The deal involved a dozen or more of several company's leaders, and would have put people like the previously mentioned J.F. Steward—who openly challenged most everything the McCormicks had to say—in business with several McCormicks and a pile of Deerings as well as miscellaneous executives from the Milwaukee, Champion, and Plano companies. Not to mention Nettie Fowler McCormick, the formidable widow of Cyrus, Sr. Nettie took a very active role in company business, more so than was expected of a woman of her time; more, in fact, than most people even dreamed possible. Only the foolish or the ignorant made major company decisions without consulting with the Harvester matriarch.

The deal was made with sweat and sleepless nights. The McCormicks reportedly didn't sleep for a week straight. When all was said and done, the Harvester companies had forged a dynasty that would rule the agricultural industry for nearly half a century.

The Formation of a Giant

Imagine that Ford, General Motors (GM), and Chrysler combined to form one company called, say, the International Automobile Company (IAC). Imagine that on that very same day, Toyota, Honda, and Nissan were bought out by an anonymous company that turned out to be the newly formed IAC. Then imagine that immediately afterward, this monstrous company began buying up steel mills, carburetor manufacturers, battery companies, and tire manufacturers. Despite the fact that the IAC would still have some competition—Saab, Porsche, BMW, Yugo, and so on—such a merger would be considered a monopoly.

The formation of the International Harvester Company was just such a merger. Literally overnight, about 85 percent of the domestic harvesting machinery manufacturers joined forces under one roof. Some of the companies claimed to be unaware of who bought them, and those that knew kept the deal close to the vest. The prestigious law firm of J. P. Morgan swung the deal and walked off with an exorbitant sum of money—rumored to be as much as $5 million—for the firm's efforts.

The principle parties in the merger were McCormick, Deering, Champion, Milwaukee, and Plano. They, along with a few smaller companies, became the International Harvester Company (IHC). The company was valued at $120 million, and controlled the market overnight.

The company did not become a single entity initially, nor could it have. As the upper echelons of power consolidated, the extremities—the dealerships—remained completely autonomous. In some towns, IHC owned as many as five separate dealerships. In most towns, there were both a McCormick and a Deering dealership. Whether the customer bought from McCormick, Deering, or a host of others, the money went into IHC coffers. This behavior got the company evicted from several states, but the money come pouring in nonetheless.

In 1912, the Sherman Anti-Trust Act was created, and IHC was at the top of the wrong list. A tremendous five-year court battle ensued, but the company managed to escape mostly intact. International had to sell off several smaller companies and consolidate dealerships, but the settlement could have been much worse for IHC. One of the key points in the defense of International was that the company wielded its tremendous power with a relatively gentle touch. Competitors were spared no quarter, but the farmer undoubtedly benefited from fair pricing and good products. Tremendous resources were devoted to research and development, making IHC one of the most innovative companies of the early part of this century. That tradition of innovation and commitment to quality allowed IHC to take control of the tractor market and hold it for nearly 50 years. Even in the waning days of the 1980s, IHC continued to be a powerful, creative force in the industry. But the dark days of the mid-1980s are many pages away. The beginning of a long, proud tradition begins with the first International tractor.

Chapter One

The Era of Innovation

The Tractor Age Begins

"If you're not failing every now and again, it's a sign you're not
doing anything very innovative."
—Woody Allen

The power on the farm at the turn of the century was the horse. The horse was versatile, economical, and performed a host of farm tasks. Working with a horse was often as back-breaking as doing it by hand, but it was the most efficient way to farm.

Mechanized power was relegated to a few specialized tasks. Lumbering steam tractors were useful for turning the belt on the threshing machine or perhaps pulling some otherwise immovable object, but not much more. Besides, they were expensive and blew up often enough to keep them out of the hands of the average farmer.

Gas-powered machines of the day were crude, foul-tempered machines, but they had their advantages over steam. To get steam up and running took from 40 minutes to an hour and a half, and required large quantities of combustibles. In the wooded north or areas near a adequate source of coal, this wasn't as much of a problem. The Great Plains were another matter, and powering threshing machines to harvest grain required hauling tons of wood or coal from far away, which made steam engines expensive to run.

While it might require a few minutes of fiddling to get a gas-powered machine up and running, such a machine was ready to go to work once it was started. Fuel was often easier to transport, if not simpler to find, than bulkier combustibles, not to mention that the fuel—gasoline, kerosene, or distillate—was relatively cheap when compared to fueling a steamer all day.

The wise, the well-read, and the gambling business folk of the time recognized that such power was going to change the farm, and were doing what they could to capitalize on the trend. Most also realized that gas tractors were the future, and it didn't take a fortune teller to see that useful, powered farm machinery had almost unlimited potential.

With IHC buying companies as frequently as traveling salesmen buy lunch, it was only natural that they would put a few dollars into a not-so-long-shot—tractors. The company's initial tractor efforts were created by purchasing a chassis and fitting it with an IHC engine .

The company put several of their best engineers to work designing and building International Harvester tractors. The IHC engineers were some of the most elite in the business, but research and development take time. International needed to get its feet wet with product.

The company already had a successful line of stationary gasoline engines. All they needed was a chassis to be in the tractor business. They found one, the Morton tractor, and began building tractors in 1906.

The company originally had its tractors built under contract, although IHC soon sold more than the Ohio Manufacturing Company's Morton plant could build, and began production in two IHC factories. Things moved rapidly from that point, with the original design—the Friction-Drive Tractor—refined and shaped into the Type A and B. The tractors were crude and difficult to use, but the experience convinced IHC that there was a future in tractors.

The ensuing few years saw a rush of development yet a low volume of sales. Sales of a few hundred units a year was considered successful, and selling less than a hundred units during a season was not uncommon.

The first decade of International Harvester tractor production was a confusing time, as development was continuous and chaotic. The company's factories were ill-equipped to build such machines, with space hard to find. In fact, some of the early 45-horsepower machines had to be built in tents because of the lack of space!

A plethora of machines which appeared quite similar but used distinctly different designs were built, and very few, if any, of them were sold. To further confuse things, two separate departments were formed and built two independent tractor lines, the Titans and Moguls. There was a lot of confusion concerning these two lines, especially on the 45-horsepower machines. If that wasn't enough, some of the machines bore several different names along the way.

The diversity of product lines and model names may have been a deliberate attempt to have unique machines for different dealerships. In *The Century of the Reaper*, C. H. McCormick states that the company intention was that the Mogul line would be sold in McCormick dealerships and the Titan line would appear in Deering dealerships. McCormick also wrote that the plan didn't work, insinuating that both lines could be sold in either of the dealerships.

Perhaps more likely is that the people brought together by the merger and subsequent flurry of acquisitions were a large, diverse group of people, even if they were part of the same organization. These people didn't necessarily share the same ideas, and were often located hundreds of miles apart. In some cases, they had been competitors one day, and engineering department bedfellows the next. .

Despite the infrastructural challenges, IHC produced reliable, well-made examples of the crude tractor of the 1910s As tractors rolled out, money rolled in, and the company invested in research and development. After only five years of building tractors, IHC was doing what it did throughout its history: building quality machines for sale and experimenting wildly behind the scenes.

Friction-Drive Tractor

Before the company could experiment, it needed to put a tractor on the market. It did so by making a deal with the Ohio Manufacturing Company, a company that owned the rights to the Morton tractor. So the first gas tractor to bear IHC markings, the Friction-Drive Tractor, was an International engine installed in a chassis produced by the Ohio Manufacturing Company.

The chassis was based on the Morton tractor, which was developed by S. S. Morton of York, Pennsylvania. Morton developed his tractor

This is a 1907 Friction-Drive Tractor. Only about 1,000 were produced. The engine ran at about 250 rpm, and top speed was 2 miles per hour.

around the turn of the century, and patented several of its designs. Morton soon joined forces with the Ohio Manufacturing Company. In 1906, IHC contracted the Ohio Manufacturing Company to combine IHC's Famous engines with the Morton tractor frame. The first Friction-Drive Tractors were produced in 1906 in very limited numbers (14 or 25, depending on which source you believe).

The first IHC tractor was a friction-drive machine powered by a one-cylinder engine. The crankcase was not sealed, and used total-loss oiling. "Total-loss" is a bit deceptive. Grease or oil cups steadily released lubricant on the bearing surfaces, keeping them more or less lubricated. Once the cups warmed, they emptied quickly. The engine was engaged by moving it back and forth on rollers. By moving the engine, a 12-inch pulley mounted on the end of the crankshaft was brought into contact with a 50-inch wheel. When the larger wheel turned, it drove the rear wheels through bull gears and pinions.

Keep in mind that the Friction-Drive Tractor weighed in at about 13,500 pounds. Friction-drive can be questionably effective on lawn mowers (some cheap ones use such a system); one can only imagine the problems of trying to move more than six tons of tractor through adverse conditions with friction-drive.

The Famous engine was from International's stationary engine line. Several ignition systems were available on early IHC engines, all of which were a bit cantankerous unless properly maintained. Early ignitions needed to be kept in tip-top condition to actually work.

Spray- or open-tank cooling kept the engine cool but used a lot of water due to evaporation. Spray-tank cooling uses a pipe with holes, a screened tank, and a catch basin at the bottom of the tank. Water is pumped through the engine's jacket and up into the pipe and out the holes, where it runs down the screen and cools. Most of it is captured and flows back into the engine, where it is pumped through the water jacket and back up the pipe. As you can imagine, water loss caused by evaporation was a problem, and cold-weather operation was a pain, because the water had to be drained out every night. In the morning, you had to fill the cooling system with hot water. The hot water helped warm the engine, but starting was still quite difficult.

The Morton frame was constructed of heavy channel iron, with several main beams connected by cross-members. This frame style was known as a channel frame, and was the most popular type of frame used on early tractors. A few used an integral frame, which meant that the engine, transmission, and/or final drive housing were stressed members of the frame. The channel frame was relatively simple to construct, but twisted under load. The low horsepower and loose tolerances of early tractors meant that a bit of chassis twisting was not much of a problem, but later machines would struggle with channel frames.

The first few Friction-Drive Tractors to roll off the line had tops made of slats with painted canvas and exhaust pipes the ended short of the top. R. W. Henderson, manager of the road engineers with the company, reported that these early machines blew so much exhaust gas, sand, and grit about that they were nearly impossible to operate. The roofs were quickly changed to corrugated tin, with elongated exhaust pipes that extended beyond the roof, which improved conditions for the operator of the tractor to some degree.

Type A

The next International tractor was still a collaborative effort, with the Ohio Manufacturing Company designing a gear drive to replace the friction-drive on the original chassis. The design was an exercise in manufacturing refinement, as the specifications from Ohio had such loose tolerances that the gears simply would not mesh. IHC engineers adapted the design and constructed the Type A gear-drive machine. The tractor used a gear-driven forward and gear-driven reverse. Note that a later Type A was also built, this one with a two-speed transmission and a gear-drive forward and friction-drive reverse.

The Type A design arrived from the Ohio Manufacturing Company in crude form, probably as a hand-built sample. C. N. Hostetter, the Superintendent of the Experimental Department, recalls that the sample did not come with drawings or specifications, and that the gears did not use a standard pitch. The first attempt to duplicate the gear-drive design resulted in a machine with gears that either could not be driven into place or simply did not touch at all. According to Hostetter, the IHC engineers conferred and decided to make an appropriate engineering drawing and simply discard the samples. Despite the fact that IHC bought the Type A design, enough of the engineering was performed in-house for the Type A to earn the IHC name!

The Type A used two friction clutches rather than a friction drive. The larger one moved the tractor forward, while a smaller one engaged an intermediate gear that put the tractor in reverse.

In 1909, the 12-horsepower, two-speed Type A was introduced. The tractor featured a gear-driven forward drive and friction-drive reverse, a Famous engine, and stub (independent) rear axles. The friction-drive reverse was an interesting choice. International said it reduced the possibility of stripping the gears by putting it in reverse while still moving forward. Whether this was actually a problem or the friction-drive reverse was cheaper and simpler to build is unknown, but many of the early tractors used a gear-drive forward and friction-drive reverse (see chart on page 26 for specific models).

Two forward speeds had obvious advantages over one, and International described the tractor as meeting the need for a "fast-moving tractor." Considering the early tractor engines ran at about 240 rpm and propelled the tractors forward at a couple of miles per hour, "fast-moving" was only relative.

Type B

The next new model was the Type B tractor, which was similar to the Type A. It used the gear reverse and forward, larger 64-inch rear drive wheels, and was available only with a 20-horsepower engine. Like the Type A, the Type B was built initially with a single-speed gear forward and reverse; a Type B Two-Speed model with a gear forward and friction-drive reverse was built later.

Both versions of the Type B were produced in low quantities, and only a few exist today. According to LeRoy Baumgardner, one of the few collectors who own one of these rare machines, the Type Bs use a different part numbering system than other International tractors and IHC logos are not found on the existing Type Bs. This indicates that Type Bs were built at the Ohio Manufacturing Company's Upper Sandusky plant. A few were built in the IHC plant in Akron, Ohio, and those would have probably been stamped with IHC logos and part numbers.

Type C

The next new model for International was the Type C. The Type C debuted with the friction-drive reverse and gear-drive forward found on the Type A and B two-speed tractors, and used a single-cylinder engine, as did all the early International tractors. The Type C used a single friction clutch rather than the dual clutches of the A and B two-speeds. The single clutch was spring-loaded to absorb the shock of sudden engagement.

The Type C was the first early International tractor to sell more than 1,000 units in a year, selling 1,787 in 1910. This tractor was one of the first to be designed by the International team of engineers, and was quite different from the Type A and B (although the appearances are similar). The tractor was developed in the fall of 1909, and was built at Akron Works for a short time, with production transferred to Milwaukee Works in 1910. About the same time, the name became the Mogul Type C, making it one of the few early tractors built at Milwaukee Works to bear the Mogul name. These tractors were available in 20- or 25-horsepower versions.

Type D Titans

The Type D Titans were introduced as a new line of International tractors. They included 20- and 25-horsepower models which were redesigns of the earlier machines, and a brand-new Titan 45-horsepower model. The 45-horsepower model was one of the first International tractors to use an engine designed specifically for a tractor.

The smaller machines were heavy-duty units, despite their low horsepower ratings. They weighed in at about 14,000 pounds, and differed from the Type Cs in their use of a live axle and larger 70-inch rear drive wheels. The tractors were sold as "Reliance" tractors for a time, and then became known as Titans.

Official production for the new 45 was authorized in May 1910, and the tractor was to be built as the 45-horsepower Reliance. It used a heavy channel frame, and was to be outfitted with the same basic design as the smaller Type D (Reliance) tractors. The tractor was to be manufactured at Milwaukee Works.

The Type D 45-horsepower design was based on the smaller machine's chassis, with an all-new two-cylinder engine with two gear-driven forward speeds and a friction reverse. The tractor was significantly larger than previous Internationals, weighing in at 18,500 pounds and standing about 16 feet long and 9 feet wide. The tractor was a radical design for its time. It used removable cylinder sleeves, crankcase compression, automobile steering, and a rear differential.

The tractor's biggest problem was hard starting. The big twins had a 9x12-inch bore and stroke, and ran at 320 rpm. To get the charge fired in the huge combustion chambers, the engine needed a strong spark. Early engines didn't provide a good spark until the engine was turning at a respectable clip, which was difficult to do with just a crank. Impulse couplings on the magneto provided a partial solution. An impulse coupling built up tension in a spring that released suddenly, snapping the magneto over at a pace brisk enough to supply a fat spark. Not all Titans even had magneto ignition, and all were still difficult to get fired up.

International provided a solution to this problem by equipping the Titan 45 with an air starter. The change took place in 1912, and could not be retrofitted to older machines as it required a host of new parts, including a new set of frame channels and a new front bolster. The number of new parts involved is evidence of how big a problem it was to start early 45s. The air starter used an on-board compressor to store air in a tank. The tank was piped to the right-hand cylinder head, and could be released to turn over the engine.

In 1912, a new version of the 45 was introduced equipped with an improved engine. This machine was quite similar to the old 45, with a different crankshaft and flywheel and a throttling governor.

This Titan 30-60 and Sanders plow are shown plowing heavy growth that had been left unplowed for 50 years.

The next big change came in 1914, when a new fuel mixer that increased power output was added. The original rating was 27 drawbar and 45 belt horsepower, although it is more typical to see the original model listed as the Titan 45 rather than the 27-45. The new model brought the ratings up to 30-60, respectively, and the Titan 30-60 was born.

Late in 1914, the 30-60 was redesigned. A fan and radiator cooling system replaced the tank cooling, and a starting engine from the Mogul replaced the compressed-air starting system. A cab was added, some of the controls were moved, and the new machine was fitted with truss rods to reduce vibration. The tractor was released for production early in 1915.

Mogul 45

About a year after the Titan 45 was authorized, the Mogul 45 was approved. Interestingly, the Mogul's specifications—horsepower, size, and weight—were nearly identical to the Titan 45s, yet the design was quite different. It is easy to see that the Titan 45 and Mogul 45 were built by

International Tractor Models & Production

Model Name	Horsepower*	Drive**	Years	Factory***	Prod #
Friction-Drive Tractor	10, 12, 15, 20	F/F	1906–09	Upper Sandusky	376
Friction-Drive Tractor	12, 15, 20	F/F	1908–10	Akron	446
Friction-Drive Tractor	NA	F/F	1908–09	Milwaukee	236
Type A Gear-Drive	12, 15, 20	G/G	1907–11	Upper Sandusky	248
Type A Gear-Drive	15, 20	G/G	1909–11	Akron	359
Type A Two-Speed	12	G/F	1909–12	Upper Sandusky	65
Type A Two-Speed	12, 15	G/F	1910–17	Akron	203
Type B Gear-Drive	20	G/G	1908–12	Upper Sandusky	255
Type B Gear-Drive	20	G/G	1910	Akron	46
Type B Two-Speed Gear-Drive	20	G/F	1910–18	Upper Sandusky	383
Mogul Type C Gear-Drive	20	G/F	1909–14	Milwaukee	2,441
Mogul Type C 25-hp	25	G/F	1911–14	Milwaukee	862
Titan Type D 20-hp	20	G/G	1910–14	Milwaukee	274
Titan Type D 25-hp	25	G/G	1910–14	Milwaukee	1,757
Titan Type D 45-hp	45, 60	G/F	1910–15	Milwaukee	1,319
Titan 30-60	60	G/F	1914–17	Milwaukee	176
Mogul 45-hp	45	G/F	1911–17	Tractor Works	2,437
Mogul Junior 25-hp	25	G/F	1911–13	Tractor Works	812
Mogul 10-20	10	G/F	1912–13	Tractor Works	85

International Tractor Models & Production (cont.)

Model Name	Horsepower*	Drive**	Years	Factory***	Prod #
Titan Convertible Road Roller	20, 25	G/G	1912–14	Milwaukee	90
Type D 18-35	35	G/G	1912–15	Milwaukee	259
Mogul 15-30	30	G/F	1913–15	Tractor Works	527
Mogul 12-25	25	G/G	1913–18	Tractor Works	1,543

Horsepower is rated by IHC; roughly equivalent to belt horsepower
**Drive refers to gear or friction-drive, and both forward and reverse are listed. For
example, "G/F" indicates a gear forward drive and a friction reverse.*
***Note that machines produced at two different factories are listed on two lines.*
Source: McCormick/IHC Archives, "Tractor Production Schedule"

Early International Tractor Production

*Early IH tractors were built and designed by several engineering and
manufacturing groups. The first were contract-built, and later designs were
developed by groups at the Milwaukee, Wisconsin plant and at Tractor Works in
Chicago, Illinois.*

---

two separate departments, with different ideas about what would work.
Some advances crossed over, as the starting engine used on the Mogul
was later used on the big Titan.

The Mogul used an opposed-twin-cylinder motor with a
9 1/2x12-inch bore and stroke that was slightly larger than the Titan's;
horsepower, however, was the same, at 45. The Mogul engine revved
slightly higher than the Titan's, running at 350 rpm.

Building the new engine caused all kinds of problems at the factory.
The size of the castings alone was problematic. The factory was set up
to produce smaller parts associated with auto buggies and reapers, so
the casting of larger items had to be contracted out. The wheels alone
were so large and heavy that a shed had to be outfitted with a large roll,
drill press, and a chain hoist. The roof of the shed had to be strengthened to handle the load of hoisting one massive wheel! The spokes and
rivets were riveted by hand or with a pneumatic rivet hammer, a time-

consuming method for even that early period. The first sheet metal tractor cab was also constructed in this shed.

In 1912, the Mogul 45's bore was increased to 10 inches and the rating was upped to 60 horsepower. The tractor was designated the Mogul 30-60, and was built until 1917.

The big 45s—the Mogul and the Titan—reflected the common belief of the time that tractors would be most useful for heavy plowing and other work requiring more muscle than a few horses. But big tractors were useful only to a limited fraction of farmers. The machines were expensive to own and operate and they were effective mainly for plowing large tracts of land and turning a belt all day long. Horses were still much more effective for most tasks, and the big machines were built and sold in very limited numbers. Although International was completely committed to building tractors by the mid-1910s, its doubtful the profits from the sales of the 45s were significant enough to attract much attention in a $120 million company.

Moguls and More

A host of smaller models were built during the early and mid-1910s, none of which enjoyed the success of later small International tractors. These included the smaller Moguls, as well as an assortment of Titans. These tractors were built in such limited production numbers that they could almost be considered experimentals. The experience gained in designing and building the machines was certainly more valuable than the revenue they generated.

The largest of the small Moguls was the Mogul Junior 25-horsepower. It used a single-cylinder engine that was basically one-half of a Mogul 45 engine. The Mogul Junior weighed 16,300 pounds.

The Mogul Junior 10-20 was also built at Tractor Works in Chicago. A total of 85 units were built in 1912 and 1913. It used a single-cylinder engine, and was the forerunner to the Mogul 15-30. Both of these models were reasonably successful. The 15-30 was discontinued in 1916, and 527 were built. The Mogul 12-25 was an opposed twin and was produced until 1918, with a total of 1,543 built.

Another one of the small tractors was the Titan 18-35. This tractor was a smaller twin-cylinder with two forward speeds and automobile-type steering. Less than 300 examples were built from 1912 to 1914.

The early small International tractors were successful enough to convince IHC that tractors were worth developing further, but they were extremely crude, heavy, and under-powered machines. Most used exposed cooling, oiling and valves, making them vulnerable to dust or dirt. The machines fared poorly in the fields, and required constant maintenance. Visionaries could see that the future of farming lay with powered equipment, but the average Joe undoubtedly had a chuckle or two at the expense of these unreliable, awkward machines as they sat wheezing, steaming, and smoking in the yard of some farmer rich or gullible enough to purchase the thing. The farm tractor's beginnings were not pretty.

The Anti-Trust Suit

In August of 1902, several of the largest agricultural companies in the United States joined to become a powerful, unsettled giant capable of dominating the farm machinery marketplace. The biggest companies to join were the McCormick Harvesting Company and the Deering Harvesting Company. Both were huge, essentially family-run businesses, and the two were savage competitors in a tough, increasingly crowded market. The merger also swallowed up a number of lesser (but still large) harvesting companies, including the Plano Harvester Company, the Warder, Bushnell, & Glessner Company (who produced Champion harvesters), D.M. Osborne & Company, and the Milwaukee Harvester Company. Approximately 85 percent of the harvesting business instantly joined forces under one corporate roof.

The architect for the deal was the J. P. Morgan law firm. George W. Perkins would take the fall when the firm and IHC ended up in court over the matter, but one would suspect Morgan to have taken a more active role than reported considering the fact that the Morgan family came away from the deal with $20 million worth of IHC stock.

The deal was arranged so all of the companies sold out to one George W. Lane, who in turn sold them to the New Jersey-based International Harvester Company. Each participating company received stock in the new $120 million company that would be controlled for 10 years by a voting trust made up of the top three shareholders: Cyrus H. McCormick, Charles Deering, and George W. Perkins.

The Morgan firm is credited with coming up with the International Harvester Company name, perhaps to emphasize that the company was formed to dominate internationally rather than domestically. But domestic control was something in which IHC excelled, and the company swallowed dozens of competitors and suppliers over the next few years.

One of the issues to debate is whether the company actually united at all. At least in the early days, an IHC logo was stamped next to each company's product logos and dealers remained independent. For 10 years, the "company" controlled several different product lines sold in different dealerships in towns across the country. In 1912, most of the lesser lines were dropped and McCormick and Deering were pretty much the only remaining dealerships. Still, there was a McCormick and a Deering dealership in almost every town, according to Cyrus McCormick, each with identical product lines but different marketing.

Several states, including Texas and Missouri, forbade IHC from doing business within state lines. The governments of these states felt that IHC was a trust, or monopoly, and competed unfairly. Why there was not more public outcry or legal action taken by the government is a mystery. IHC would later be accused of bribing legislators, dodging taxes, receiving kickbacks from steel manufacturers, and purchasing competitors solely for the purpose of dismantling them, but none of these charges surfaced until 1910, eight years after the company was formed. Ron Chernow in *The House of Morgan* wrote that the reason for the reprieve was that J.P. Morgan and the trust pressured President Roosevelt to restrain the justice department from prosecuting.

The trust-busting President Taft was not as easily dissuaded, and he filed suit against IHC in 1911. The result was a trial drawn out by IHC promises to correct its errors, and then by IHC proposals that were rejected time and again by government officials. A protracted, sordid lawsuit took place, amid which IHC fought bitterly to keep both McCormick and Deering under the same roof. The company prevailed, and the end result was that IHC had to sell off the Champion, Osborne, and Milwaukee lines and unify its product line. International was also forced to market all products under the name "International."

In 1917, the Titan 15-30 became the International 15-30, as did a handful of other IHC tractors. Renaming the tractors was not really the issue of importance, as all that was required was essentially a new set of decals. Besides, the International name didn't really stick, and the company continued to market its tractors under a variety of names—Titan, Mogul, McCormick-Deering, and Farmall.

The real issue that affected IHC was the requirement to consolidate its dealerships. As mentioned above, there were often several IHC dealerships operating in the same town under a different name. In some cases, there were as many as five dealerships in a single town. As a result of the order, IHC had to close hundreds of dealerships and consolidate these into one per town. This didn't cost IHC quite half of its dealerships, but it was close. According to Cyrus McCormick in *The Century of the Reaper*, IHC had 21,800 dealers in 1917, and cut back to 13,860 by 1919.

While McCormick stated that the court's decree cost IHC some business volume, he pointed out there were advantages. The company combined the harvesters, creating the McCormick-Deering line. The new machines combined the best features of both lines. McCormick stated that salesmen were pleased to be selling a better product, that customer service improved, and that the resultant joint experimental department had increased resources and improved focus.

During the trial, IHC was able to flex enough might to cool President Taft's public declarations of war against the company and defuse the anti-trust suit. The company called the result a "moral victory," because the government could not prove that IHC had abused its market dominance. The decision was more than that. While perhaps not a total victory, as the company was broken up a bit, IHC went on to control the tractor market for 40 years and become one of the dominant tractor manufacturers of the twentieth century.

Despite the lawsuit, Cyrus McCormick was optimistic when he spoke at the seventh annual Harvester Club Dinner in February of 1917. He said, "Whatever the decision, we know there is a large future ahead for this organization, because an organization like this cannot be turned aside from doing a great deal of good in the world."

The company came out of the affair in good shape, but there is really little question that the formation IHC created a monopoly. Most of the competition was absorbed, independent dealers competed for the customers' dollar under the umbrella of one organization, and the sales to Charles Lane and huge payment to the J. P. Morgan law firm was a convoluted (and mostly successful) attempt to dodge the law.

Some of the anti-trust legislation of these days came under fire by Alan Greenspan and others, and the end results of the conglomeration were not necessarily detrimental to the consumer. That was probably one of the most compelling reasons for the government to leave IHC nearly intact. The competition suffered, but the consumer did not.

As time went on, the company continued to prosper, but the shadow of the anti-trust suit lurked in the wings. A long-time IHC employee who was hired in the 1940s once remarked that the lawsuit was a topic that was simply not discussed, even through the 1960s and 1970s.

Throughout the 80-plus years of IHC, the company was one of the most innovative in the industry. Perhaps, just perhaps, some of that commitment to innovation and experimentation stemmed from a ghost of a feeling in the back of the collective corporate mind that IHC owed someone something.

The company repaid any real or imagined debt with a long history of advances that made the farmer's life easier and more efficient. Whatever the Harvester trust cost American business of the early 1900s, IHC repaid—plus interest and dividends—several times over in the ensuing eight decades.

The First Small Tractors

One of the trends that characterized the flurry of tractor development of the mid- to late-1910s was the attempt to build a successful small tractor. Dozens of tractor manufacturers sprung up overnight to attempt to satisfy this perceived need. And most of the agricultural world knew that the company that built a tractor small enough to be used for everyday tasks and powerful and reliable enough to perform them with ease would make a bundle. The only question was who and when (a simplistic answer to that being Henry Ford in 1917, but that story is reserved for a bit later).

Although International had produced several machines with less than 20 horsepower, all of the early models had been nearly as large and heavy as the more powerful machines. Despite this, International had been plugging away at light tractors from day one, with strange little machines crabbing around the yard since as early as 1905.

With the Titan 15-30, the Mogul 8-16, and the Titan 10-20, International found success with smaller tractors. These machines were relatively light at less than five tons, and reasonably priced. The tractors were fairly reliable, and well-built by the standards

Titan and International 12-25/15-30 Tractor Production	
Year	Production
1914	13
1915	196
1915	511
1916	60
1917	376*
1918	1,285
1919	1,652
1920	1,068
1921	821
Total Prod.	5,982

*Name changed from Titan 15-30 to International 15-30 in 1917.
Source: McCormick/IHC Archives, "Tractor Production Schedule"

of the day. They reflected the beginning of the tractor's transition from a heavy, limited accessory to a vital tool on the farm.

Titan 15-30

These tractors were known as "light tractors," that weighed a sprightly 9,300 pounds, giving it an unprecedented power-to-weight ratio of 310 pounds per horsepower.

The tractor debuted as the Titan 12-25 late in 1914, sporting a four-cylinder engine. Shortly after the introduction, the rating was upped to 15-30. Consequently, the Titan 12-25 and 15-30 are one and the same. The 15-30 was plagued with name changes, as the Titan name was dropped in 1917 as the result of an anti-trust suit. In 1917, IHC was mandated to name all of its products "International." So, the Titan 15-30 became the International 15-30. Perhaps the company's distaste for the mandate is why IHC tractors continued to be marketed by a wide variety of names—Titan, Mogul, McCormick-Deering, Farmall—until the 1960s, when they became more widely known as simply "International" tractors. The Titan 15-30 remained in production until 1921, when it was displaced by the McCormick-Deering 15-30.

Mogul 8-16

Introduced in 1914, the 8-16 was a single-cylinder tractor with a distinctive curved frame that rose up over the front wheels, allowing a tight 20-foot turning radius. The tractor was quite narrow and smaller than the typical International tractor. The engine was a horizontal single-cylinder that was hopper-cooled and fired by magneto ignition. A planetary gear transmission powered the rear wheels through a single left-side chain final drive. The tractor had one forward and one reverse gear.

In 1915, IHC sold 5,111 machines, the bulk of IHC's tractor production that year. Sales numbers continued to be strong into

Mogul 8-16 & 10-20 Production	
Year	Production
Mogul 8-16	
1914	20
1915	5,111
1916	8,269
1917	665
Total Prod.	14,065
Mogul 10-20	
1916	25
1917	5,338
1918	3,146
1919	476
Total Prod.	8,985

Source: McCormick/IHC Archives, "Tractor Production Schedule"

1916, with over 8,000 sold. In 1917, the tractor was updated with a larger cylinder bore and a two-speed sliding gear transmission and dubbed the Mogul 10-20. Sales continued to be strong that year, and the tractor was discontinued in 1919, most likely due to the popularity of the Titan 10-20.

Titan 10-20

The Titan 10-20 was the first great success of the IHC tractor line. The size was right, the timing was right, and—thanks to Henry Ford—the price became very attractive. This tractor was not especially innovative, with the standard channel frame, chain final drive, magneto ignition, and two-forward-speed gear-drive transmission. The equipment was hardly earth-shattering, but the package combined all the current technology in a simple, well-built, and relatively reliable package.

Development of the Titan 10-20 began in the Fall of 1914. By April 1915, the first prototype was plowing fields near Milwaukee, and soon after two more prototypes were sent to Texas for testing. In the summer

The Mogul 8-16 was the machine that convinced IHC of the viability of the small tractor. It sold briskly, quickly outpacing the larger machines that had received the lion's share of the company's efforts.

of 1915, seven more were built and sold. The first of these created a sensation when it successfully completed a 60-hour non-stop plowing demonstration at Carlinfield, Illinois. The Titan was authorized for regular production in 1916, and it was produced at Milwaukee Works.

The Titan 10-20 had a larger-displacement two-cylinder engine, a bit more weight, and more pulling power than the Mogul 8-16. The 10-20 was rated as a three-plow tractor, which was considered the ideal size. Farmers responded favorably to the tractor, and it sold well until 1922, the last of year of production. It's fairly amazing that the tractor sold at all in 1922, as the McCormick-Deering 15-30, a tractor that was light years ahead of the Titan, was introduced in 1921.

Titan 10-20 Production

Year	Production
1915	7
1916	2,246
1917	9,044
1918	17,675
1919	17,234
1920	21,503
1921	7,729
1922	2,925
Total Prod.	78,363

Source: McCormick/IHC Archives, "Tractor Production Schedule"

The Titan 10-20's engine rested in the chassis horizontally, with the two cylinders parallel. Oiling was total-loss, meaning that brass and glass oil containers dripped a slow but steady flow of oil onto bearings. Carburetion was by a fuel mixer until 1921, when an Ensign JTW carburetor was used. The Titan was started on gasoline, and then switched over to kerosene, and IHC promoted kerosene as the fuel of choice for farming. An IHC advertisement of 1919 read, "Kerosene is the practical tractor fuel. Don't let yourself be led away from this fact. Gasoline is an unwarranted extravagance." At the time, gasoline was significantly more expensive than kerosene.

The twin-cylinder engine was cooled by a thermosyphon system, which used the engine's heat to circulate coolant from the large water tank to the cylinder head jacket. The water tank system was not the most efficient, but was an improvement over the old hopper-cooled Titans. The engine and transmission were mounted on a steel frame. The tractor weighed in at about 9,000 pounds, which was reasonable at the time.

Power was transmitted to the rear wheels by dual roller chains. Two forward speeds were provided, as well as reverse. The tractors featured small, narrow rear wheel fenders until 1919, when larger full-coverage fenders and an operator's platform became standard.

World War I ended in 1918, but the Titan 10-20 truly went to battle

The Titan 10-20 was the first IHC tractor built that sold more than 10,000 units in a year. The machine was used as the main weapon in the battle against the Fordson, and went through a number of price cuts.

in 1921. At that time the Fordson was dominating the tractor market, and IHC slashed prices on the Titan to try and compete with the cheap, technologically advanced Fordson. The price for the Titan 10-20 was cut from $1,250 to $1,000 in March of 1921. The price was cut to $900 in October of 1921, and slashed to $700 in February of 1922 when Henry Ford cut Fordson prices to $395, a price that he admitted was less than his cost.

The Titan 10-20 gave IHC inklings of the possibilities of the power farming market, and opened farmers' eyes to the idea that tractor farming could be efficient and economical even on smaller farms. It was introduced into an era that saw tractor sales growing in almost unbelievable bounds, rising from 21,000 tractors sold in 1915 to more than 200,000 in 1920.

In those fast times, the Titan 10-20 was a bit of a staid player, although it performed adequately for thousands of farmers. New technology was on the way, and the Titan would turn from state-of-the-art to obsolete in less than a decade.

Chapter Two

Design By Committee

The International 8-16 and the McCormick–Deerings

"It is impossible to say whether the tractors
we are selling today will be out-of-date in four or six years,
but it seems quite probable that they will be."
–International Harvester Company, 1924

The late 1910s were a wild time for tractor development and a challenging time for International. Technology was changing at an unprecedented speed, and new manufacturers were forming overnight. The feverish growth of the times was similar to the computer revolution, with machines developing so rapidly that each model became obsolete in a few years. The surging technology transformed the tractor from a wheezing beast to an effective tool, and the sales numbers that everyone anticipated emerged. But the combination of the anti-trust suit settlement and the success of the Fordson made the late 1910s and early 1920s a critical, difficult time for International.

Beginning with the International 8-16, IHC's engineering and development work began to bring its tractors above the rest of the industry.

More than a decade of experimentation bore fruit with new technologies and refined, effective machinery. The International 8-16 pioneered the power take-off, and was used as a test mule for most of the company's new ideas. The 8-16 also served as the development platform for the McCormick-Deering 15-30 and 10-20, which set new standards for durability and ease of service.

The International 8-16's reception was spoiled by the U.S. Government and Henry Ford. The government forced International to consolidate their dealership network. After the merger, IHC at times had three or four locations in one town; the settlement required the company to close up all but one.

Henry Ford would offer an even more difficult challenge. His new tractor, the Fordson, appeared in 1917 and quickly devoured the market. It was light and cheap and backed by a man who was practically a national hero. Despite rising tractor sales, the International Harvester Company was in a life-and-death battle just to stay alive. The company's top weapon should have been the International 8-16, but production woes kept it from reaching the dealerships in sufficient quantities to meet demand. As a result, the stodgy but well-built Titan 10-20 was brought to the front line.

International Harvester introduced two new models into that growing market, the McCormick-Deering 10-20 and 15-30. Both were designed to beat the Fordson at its own game, and the result was a couple of machines that were far superior to the Fordson. These tractors were designed with production-line manufacturing in mind, and used the latest technology—enclosed gear final drive, replaceable cylinder sleeves, radiator cooling, carburetors, and more. Although the Farmall is commonly credited as being the tractor that knocked the Fordson out of the domestic market late in the 1920s, the McCormick-Deerings put the Fordson on the ropes.

The McCormick-Deerings also pioneered IHC's first serious attempt to sell industrial tractors. The line would become very successful, as tractors were well-suited to industry tasks, and converting the tractors for such use was relatively simple and inexpensive.

The developmental platform for the McCormick-Deerings was the International 8-16, an innovative tractor that, like Rodney Dangerfield, never got much respect.

International 8-16

The International 8-16 was a radical departure for International, with a sleek, streamlined look, a vertical four-cylinder engine, and three forward speeds. It served as a developmental mule for everything from the power take-off to four-wheel-drive to experiments with rubber tires. The International 8-16 was also the first International to be built on a production line.

The channel-frame, chain-drive 8-16 was not the tractor Harvester hoped for, as production difficulties kept it from meeting early demand. Once the production line was in place and the 8-16 could be produced in quantity, the tractor's design was dated and the new McCormick-Deering models were being rolled out.

The 8-16 began life in 1914 and, after a couple years of testing, was put into production as the Mogul 8-16 four-cylinder in August 1916. The name was changed to the International 8-16 in 1917.

The 8-16 used several different engines during its production span, all of them being four-cylinder units that ran at about 1,000 rpm. The transmission was a three-speed unit, and final drive was by chain. A high-tension magneto supplied the spark, and a radiator and fan kept things cool, with coolant circulated by thermosyphon.

Engine Troubles

A variety of glitches kept it from being produced in quantity until 1918. One of the problems was the engine, or engines. Several different engines were used in production 8-16s, resulting in three different serial number series. The first engine had problems with insufficient lubrication. It was replaced, and a new set of serial numbers were assigned. The second engine was apparently under-powered, as it was replaced with a more powerful unit, and a third set of serial numbers were assigned.

One of the engines emerged from the factory with a transmission case that had the disturbing tendency of cracking. The cases used a tube that ran from the transmission to the rear drive chains, cleverly lubricating the chains. Not so cleverly, it introduced cold air into the hot oil of the transmission. If conditions were right, the transmission case could be cracked. A change order replaced the tube with a chain oiler, which was hurriedly retrofitted to the tractors in the field.

Changing engines during production was expensive and raised all kinds of havoc with company balance sheets. Some kind of exceptional force is required to make executives sign off on these kind of decisions, and Henry Ford was undoubtedly that force, since the International 8-16 could have been—and was probably intended to be—IHC's answer to the Fordson.

The 8-16 was small and light, like the Fordson, and was reasonably priced. The 8-16 weighed about 600 pounds more than the Fordson, and both used four-cylinder engines with a 4x5-inch bore and stroke that ran at 1,000 rpm. The Fordson was rated as a 10-20 and the International as an 8-16, but actual output and pulling power were comparable, according to tests performed by the Agricultural Engineering Department at Ohio State University. Features like removable cylinders, a multiple-disc dry clutch, and a sliding gear transmission made the International a better-built and longer-lived machine as well.

The most significant differences between the 8-16 and the Fordson were the retail price and the manufacturer's ability to produce enough

The International 8-16 had three forward and one reverse speeds, and a vertical four-cylinder engine with removable cylinder sleeves. The tractor was plagued with engine problems, and three different powerplants were used in the model during its lifespan.

machines to meet demand. From the farmer's perspective, the Ford was cheaper, rated for more horsepower, and available. Factor in the Henry Ford name, and it's evident why farmers were willing to ignore the Fordson's weaknesses and sign on the dotted line.

The Fordson had several shortcomings, but the biggest problem was deadly. The short wheelbase, light weight, and worm-gear final drive made the Fordson flip over backwards suddenly under heavy, sudden loads. Also, the worm-gear final drive heated up the operator's posterior something fierce, and the exhaust note assaulted the ears. Despite this, it was cheap and Ford was set up to build more than 100,000 a year.

Production Problems

The International 8-16's relatively weak sales were certainly linked to the engine difficulties as well as the manufacturing glitches, price restructuring, and engineering changes. All sorts of problems kept the 8-16 from reaching the sales floor in sufficient volume, and the delays led to in-house skirmishes between manufacturing, sales, and engineering. An IHC memo from executive Alexander Legge discusses the problems with open frustration:

"It is now three years since we commenced selling this machine [the International 8-16]. During the first year, I think the delays in progress might in the main be charged to the engineering department, owing to changes in design. During the second year, we might make it fifty-fifty, or perhaps with justice charge a considerable portion of it to the abnormal war conditions, as a year ago now we were arranging to devote a considerable portion of Tractor Works manufacturing capacity to government work, making it impossible to make progress on a regular line. However, it will soon be a year since the Armistice was signed, during which period there has been nothing to prevent our going forward with increased production and improved practice."

The memo continues, discussing converting the 8-16 from a chain final drive to a gear final drive. Company documentation shows that a gear-drive 8-16 was considered, but never built. Legge also wrote:

"The tractor may not be the ultimate limit of perfection that we hope to achieve in time, but it has gone through the year with surprisingly few complaints. Dealers who have been fortunate enough to secure a substantial number of them during this season make the statement that very little experting or service attention has been required. In any event, it is good enough so that I feel certain we could be selling a hundred a day if we were making them, yet our product is about one-fourth that number."

A follow-up memo came from company president Cyrus McCormick, who supported Legge's assertions that the onus was on manufacturing to produce more tractors. He balanced this view by questioning whether it was worth sinking tremendous sums of money into revamping the International 8-16, when its longevity was in question due to a new design (presumably the McCormick-Deering 15-30 and 10-20) that was in the works.

"What will happen . . . cannot be foretold, as we have reached no conclusion as to whether we will continue to manufacture a frame-type tractor or go over to the unit-type. I would like nothing better than to undertake a tractor program of great volume. The present 8-16 is a wonderful tractor, but will it last long enough to equip, say, on a basis of five years of unchanged production?"

Mass Production of the 8-16

McCormick's uncertainty was reflected in company actions. International 8-16s were originally handbuilt. Henry Ford showed the world how well a continuous, moving production line could work with the early Ford cars, and the rest of the industry scrambled to follow. While a mass production line increased capacity and decreased costs, the initial investment required was prodigious.

Part of IHC's plan for the new McCormick-Deering tractors was to build them on a production line. Interestingly, the International 8-16 was also built on a production line, at least for a short time. Photos from *Evolution of A Tractor*, published by Stemgas, clearly show 8-16 engines and chain-drive machines being assembled on a production line.

Creating a production line for this dated machine wouldn't make a lot of sense, as the company was in the process of phasing in new machines. It is likely that the company temporarily built the International 8-16 on the new production line with the intention of converting the line to produce the new McCormick-Deering machines. In this way, the teething problems of running a production line could be ironed out before trying to get a brand new model out the door as well.

To add an interesting and perhaps confusing twist to the story, early versions of the McCormick-Deering 10-20 were known as the International 8-16 Gear Drive Tractor in company records.

When the 8-16 could have been selling exorbitantly, the production facilities did not exist. By the time it was feasible to step up production, the International 8-16's time had passed, both from a market and a company standpoint, and the tractor was more or less abandoned.

Another aspect to the 8-16 was the development of the power take-off. Sold in conjunction with specially-designed implements, an 8-16 could increase efficiency in the fields in an unprecedented manner. It is believed to be the first American tractor sold with a practical rear power take-off.

A crude form of a rubber tire was also tested on the International 8-16. This was hardly the first use of rubber tires on a tractor, as the Rubber-Tired Steamer was built in 1871 with rubber-coated steel wheels. The 8-16 used a similar type of tire, as International engineer L.B. Sperry described it in *The Agricultural Tractor 1855–1950*: "The front wheels were fitted with solid-section rubber tires; the rear wheels with solid-section blocks molded into metal detachable lugs and the blocks were made by the Firestone Tire & Rubber Company."

The tires didn't work well on pavement, and the rubber peeled away from the steel lugs, so the idea never made it past the experimental stage.

Like the Titan 10-20, the International 8-16 was used as an agent in the price war against Ford. Prices of the 8-16 were cut steadily in the early 1920s, from $1,000 in March 1921 to $670 in February 1922. Cyrus McCormick III wrote that the cut to $670 came during a phone call in which Alexander Legge was told that Henry Ford cut the price of the Fordson below cost. Legge supposedly issued a vehement response, stating that International would meet the price. International certainly needed to respond to Ford's challenge, but the $670 price on the 8-16 probably reflects the company's desire to clear out old inventory to make room for the new McCormick-Deerings as much as it did the desire to compete with Ford.

The International 8-16 was a hallmark vehicle in many ways. Its full-coverage body work, three-forward-speed transmission, and smooth-running four-cylinder engine were all signs of things to come.

Development of the Power Take-Off

By the time the International 8-16 reached the market, the idea for using the tractor engine's power to operate implements had been around for some time. The first clearly recorded example was Aveling & Porter's steam reaper, which appeared in the December 19, 1885, issue of *Farm Implement News*. The reaper was pushed in front of the steam tractor, with a chain drive running from the steam engine's flywheel to the reaper. The machine was displayed at the Paris Exposition as early as 1878.

The next reference to a power take-off (PTO) appeared on a tractor built by Frenchman Albert Gougis. Made in 1905, this tractor was a home-built machine. Gougis ran a chain from the crankshaft of the engine to a shaft that connected to a McCormick binder. Several universal joints handled flex on the shaft, and the mechanism could be engaged separately from the clutch. In this way, the tractor pulled the binder and powered it. Gougis developed the binder to help farmers save downed grain, and it reportedly worked well for this purpose.

It has been speculated that the idea for the power take-off was sparked by Gougis' device, which was spotted during an International executive's trip to France. A photo in February 12, 1931, issue of *Farm Implement News* shows Gougis operating a McCormick binder with a home-built tractor and crude PTO. Ironically, the design of the home-built tractor bears a remarkable resemblance to the Farmall Regular. The photo was taken about 1906, and it can be speculated that International saw more possibilities in the ideas of Gougis than just a PTO.

Engineer Bert R. Benjamin proposed developing PTO drives for the Titan 10-20 as early as 1917. He reported encountering a farmer who owned several International powered binders, a Titan 10-20, a team of horses, and a Happy Farmer tractor powered by an 8-horsepower Cushman engine. Benjamin pointed out that the farmer could afford to purchase several more Titan tractors if they were equipped with a PTO, allowing the farmer to dispense with the expensive and often inefficient powered binders. Benjamin's proposal was received with favor, and a PTO was developed to power the cutter bar and sweep rake lift of the Motor Cultivator. The Motor Cultivator, a specialized machine designed purely for cultivation, failed miserably, but the engineering work opened the door for the Farmall, a tractor capable of cultivation.

In 1919, IHC began experimenting with PTO-equipped International 8-16s, and in 1921, the PTO became available as special order equipment on the International 15-30. The new McCormick-Deering 10-20 and 15-30 also appeared in 1921, and the PTO was available on both machines. The Farmall used a PTO as early as 1922, and it became an integral part of the tractor's design.

Despite the fact that the PTO was available in the early 1920s, it wasn't until later in the decade that it was widely known. The rice crop of 1925, imperiled by a wet season in Arkansas, Louisiana, and Texas, was saved by PTO-driven implements and the agricultural world woke up to the fact that the PTO had arrived.

International 8-16 Production	
Year	Production
1917	38
1918	3,162
1919	7,571
1920	5,848
1921	9,013
1922	7,506
Total Prod.	33,138

Source: McCormick/IHC Archives, "Tractor Production Schedule"

The next hurdle was standardization of PTO shaft rotation speeds, coupling size, and splining. Under the auspices of the American Society of Agricultural Engineering (ASAE), representatives from tractor manufacturing companies met to attempt to standardize the PTO coupling so that tractors and implements from different manufacturers would be interchangeable. It took about five years to become reality, but by 1931 ASAE had mediated a standard for the PTO and it was being applied to any imaginable implement. The PTO continued to develop with tractors, eventually shaking out to two standard speeds, 540 and 1,000 rpm. Later tractors had two PTO couplings, one for each speed.

Changing Times

The tractors of the 1910s and 1920s were developing as quickly as the personal computer of the 1980s and 1990s. Farmers were advised to expect to get five years out of their tractors before they were obsolete; in reality, it was often much less than that. Mind you, the five-year-old tractors still worked as well as when purchased; they simply were no longer efficient or compatible.

Even the overlap was remarkable. The McCormick-Deering 15-30 sat in the showroom alongside the International 15-30. The two were generations apart, with the International essentially a refinement of the original Titans and the McCormick-Deering a rough form of the modern tractor. The International still used total-loss oiling for most bearings, a fuel mixer rather than a carburetor, contracting-band clutch, and chain final drive. The tractor was obscenely heavy, complex to operate, and just turning them around was an aerobic endeavor.

The McCormick-Deering tractors were a departure for IHC. They were the tractors some felt would carry the company to the next level. Although the International 8-16 and Titan 10-20 fought the battle against the Fordson, the McCormick-Deerings were IHC's technologicallly superior weapon. International did not attempt to compete with the Fordson on price. Instead, the company built a machine that blew the

Fordson out of the water with quality. The McCormick-Deerings were long-lived, with roller bearings and rugged construction. They could be serviced easily, due to unit construction that allowed components to be removed easily for repair or replacement. The engine used removable cylinder sleeves, which meant that it could be rebuilt simply, and the final gear drive lasted longer and worked better than the Fordson's worm gear drive. The end result was that the McCormick-Deerings took the advantages of the Fordson—the integral frame and line production— and assembled them with into a high-quality package that was reasonably priced and, with care, would outlive the farmer.

The McCormick-Deerings were more than just high-quality machines for International. They represented the future of the company. Ford was hammering International on the sales front, and knocking prices down to bankruptcy-inducing levels. All around IHC, tractor manufacturers were going belly uptrying to compete with the Fordson. International could not hope to compete with the Titan 10-20 or the International 8-16. Both were decent machines, but the prices were cut to the point that profits were minimal, and neither machine was spectacular enough to draw attention away from the Fordson's attractive price.

IHC was desperately seeking the next great leap, which it needed to survive. The first thought was to convert the 8-16 to a gear-drive machine, but the twisting of the channel frame wouldn't allow the tight tolerances necessary for a gear final drive. In addition, the addition of gear drive still didn't bring the 8-16 up to the level that International management felt was necessary.

What the company needed was a high-quality machine that could be built in quantity. A mass production line and a new type of frame were the answers. To do both required complete retooling and the construction of a production line. The investment required was great, but the cost of lagging behind was greater. The McCormick-Deerings were given top priority, and carried the hope for the company's future.

Cyrus McCormick III had this to say: "Two new improved models, the McCormick-Deering 10-20 and 15-30, were introduced which summed up the entire story of Harvester's tractor experience. Millions of dollars were poured into the modern type of labor-saving manufacturing equipment. Production costs were slashed by the means of efficiency gained through elimination of wasted effort."

What McCormick was referring to was modular construction, which simply meant that component groups would be assembled individually.

The entire tractor would be put together somewhere else, with each component group simply bolted on. Rather than installing the clutch bit by bit onto the engine, the clutch was assembled as a unit and then slid into place when the engine and transmission were mated. This speeded up assembly a bit and, more importantly for the customer, simplified maintenance. Time is money on the assembly floor, and the result was a lower cost to the farmer and higher profit margin for IHC, as well as the ability to meet high-volume demand.

The other key aspect of the McCormick-Deerings was the integral frame, which was stronger and better sealed from dirt than the old channel frames of previous models. Keep in mind that what International called an integral frame was not the same as another company's concept of it. The term integral frame can refer to a frame that uses the engine and transmission housing as a stressed member of the frame. The Farmall Letter Series tractors used that type of integral frame.

The integral frame used on the McCormick-Deering 10-20 and 15-30 was simply a big cast iron tub that housed the engine, transmission, and final drive. The axles were bolted to this tub, and the basic equipment was in place. The design was very strong, and sealed bearing surfaces from dust and dirt.

The company believed the integral frame and unit construction, combined with the superior technology of the McCormick-Deerings, made the 10-20 and 15-30 the tractors of the future for IHC. The Farmall threw a huge wrench into that plan, as the efforts of Bert R. Benjamin produced an innovative product that used more traditional construction and a channel frame. The company had to back the obvious winner, but modular construction and an integral frame (of sorts) would reappear on the Letter Series tractors, when Benjamin's innovative tricycle design would be merged with the production methods of the 10-20 and 15-30.

Implement and Tractor columnist Elmer Baker compared the McCormick-Deerings to the Farmall by the methods they were developed. The Farmall was the result of one man's vision, while the McCormick-Deerings were designed and developed by a group. Baker wrote, "What might be called committee designing in the Harvester company at the time was evident in the International 10-20 and 15-30. They were assembly-line machines made to automotive standards the way the original Fordson had been . . .The point is that the 10-20 and 15-30 were simply good tractors—pattern tractors in fact—but they were not revolutionary."

Henry's Call to Arms

The First Round of the Tractor Wars

One of the major challenges faced by the International Harvester Company was the entrance of Henry Ford and his Fordson tractor in 1917. The Fordson was light, reasonably powerful, economical and sold by a marketing genius through his extensive network of car dealerships. Ford soon owned 75 percent of the market, and sold more Fordsons in 1919 alone than most tractor manufacturers had in the past decade.

The key to the Fordson's success was corporate might coupled with economical production. The integral frame incorporated the heavy-duty engine, transmission, and rear drive housings as part of the frame. The frame was stronger than the channel frames of the day, and the entire tractor was designed for modular construction and could be assembled quickly on a factory line. Ford could crank out hundreds of machines per day for dramatically lower cost than the competition, and was soon doing just that.

Ford had very deep corporate pockets and a zealous passion to bring power to the farm. He was willing to cut his margins to nearly nothing in order to sell his tractors.

The tractor industry had to respond to survive, and most manufacturers lowered prices in an attempt to gain some of the market that Ford had created. The low prices drove industry-wide sales, and tractor use on the farm grew dramatically over the next five years.

Ford sold even more machines by making an arrangement to distribute his Fordson tractor through government agencies as a war measure. Cyrus McCormick wrote:

> "To have convinced worried statesmen and the public that the tractor was a new device twelve years after many tractor builders had attained large production, and that his particular make would prove to be the one solution to the knotty problem of food production, was a supreme feat of salesmanship."

Ford Versus International

Year	Tractor Prod.	No. of Trac. Companies	Interational		Ford	
			Prod.	Share	Prod.	Share
1915	21,000	61	5,841	28%	-	-
1916	29,670	114	11,571	39%	-	-
1917	62,742	124	16,101	26%	259	0.4%
1918	132,697	142	25,269	19%	34,000	26%
1919	164,590	164	26,933	16%	54,000	33%
1920	203,207	166	28,419	14%	58,000	29%
1921	68,029	186	17,762	26%	13,000	19%
1922	98,794	116	11,781	12%	62,000	63%
1923	131,908	93	12,026	9%	100,000	76%
1924	116,838	64	18,749	16%	78,000	67%
1925	164,097	58	32,588	20%	100,000	61%
1926	178,074	69	50,900	29%	NA	NA
1927	194,913	61	55,727	29%	NA	NA
1928	171,469	51	94,148	55%	40,000	23%

This chart shows how Ford and International compared in total production and market share. Note that International dominated the market in 1916, then took an almost immediate back seat to Henry Ford's little machine when it appeared in 1918. Note that 1921 saw Harvester pull ahead of Ford, which may have incited Henry to drop the price of his Fordson below cost. The ensuing price war had Ford controlling the market, but International was able to outlast Ford and take control by 1928, when the Fordson was pulled from the domestic market.

While the lowered prices and increased competition brought good things to the farmer, it devastated the industry, and IHC was no exception. Almost overnight, IHC went from owning the market, controlling 39 percent in 1916, to half of that in 1918 (see chart). The Fordson was squeezing IHC, and it also had a catalyzing effect on new manufacturers eager to get a slice of the tractor market pie. The number of manufacturers peaked in 1921 at 186, an incredible number considering only a handful control the market today, and only John Deere survived intact.

But 1921 was not a good year to be starting a new tractor company. International was fighting for sales and dropped prices several times that year. They also initiated field demonstrations designed to showcase IHC tractors and convince buyers that IHC tractors were worth the extra money. The weapons were the Titan 10-20 and International 8-16, and it seems they were up to the task of taking on the Ford, as IHC regained a larger share of the sales that year, gaining 12 percent in market share. But Ford wasn't finished.

Ford responded, turning his stranglehold into a death grip by dropping the price of the Fordson to $395, a price he admitted was below cost. IHC cut its prices, as well, selling Titan 10-20s for $700 and International 8-16s for $670. The company also took the war to the fields, challenging Fordsons to plowing contests. In *Century of the Reaper*, Cyrus McCormick had this to say about the tractor wars:

> "A Harvester challenge rang through the land. Everywhere any single Ford sale was rumored, the Harvester dealer dared the Ford representative to a contest. No prizes were offered, no jury awarded merit to one or another contestant. No quarter was given and none was asked. Grimly the protagonists struggled, fiercely they battled for each sale. The reaper war was being refought with new weapons."

In 1922, IHC was not especially successful, losing market share in a market where a buyer could bring home a Fordson for less than $400. The rest of the industry fared worse. Seventy tractor companies failed in 1922, and 13 more dropped out the next year. In a three-year span, the number of tractor manufacturers had been cut in half.

By 1926, IHC had its new McCormick-Deering tractors on the market and was regaining the dominance of the past. In 1928, Ford bowed out and IHC's new unit-construction McCormick-Deerings helped IHC take control of the market; 55 percent of all tractors sold that year bore the IHC logo.

In 1931, IHC's market strength led Cyrus McCormick to write:

> "When the tractor war was over, the farmers of the world appreciated beyond a shadow of doubt that they would best serve themselves by providing their farms with a tractor rugged enough to resist the shocks of farm use and powerful enough to do all of their work. They knew that there can be no such thing as a good cheap tractor."

International maintained a dominant market position until 1939, when the introduction of a stunning line of new Farmalls was spoiled by a handshake between Henry Ford and Harry Ferguson that brought the Ford tractor back to the United States.

This is not to say that the McCormick-Deering 10-20 and 15-30 were not important tractors. They were, in fact, the state-of-the-art in the early 1920s. They were intelligently engineered and well-built, with quality components used throughout. They were light years ahead of the Titan 10-20, which was considered a good machine as late as 1920. Four years later, the McCormick-Deerings made the Titan obsolete.

McCormick-Deering 15-30

The 15-30 was the first of the McCormick-Deerings to appear. The tractor was a replacement for the International 15-30, with the open field and heavier jobs in mind. The McCormick-Deering was a leap to the next level for tractors, and had features like ball bearings throughout the engine and transmission, decent carburetion, more reliable ignition, and a lower center of gravity. Additions like the power take-off added immensely to the tractor's utility, and it was one of the first tractors that retained its usefulness for a decade or more rather than just a few years.

The two most significant advances on the new 15-30 were the gear final drive and integral frame. As mentioned earlier, International's idea of an integral frame was a large bathtub-shaped frame that housed the transmission and final drive; the engine housing was a separate piece that bolted on. No matter, International's integral frame was many times stronger than the channel frame design, and was sealed to keep dirt out of the engine and drive assembly.

The integral frame also allowed the use of the 15-30's second key advance, gear final drive. Most previous IHC tractors used roller chain final drive. Some were enclosed, but most were open to dirt and grime. The chains needed occasional adjustment, and tension was crucial. Too much and you would wear out the chain and sprockets prematurely; too little, and the chain would jump the sprockets. Oiling was also a black science. Lubrication naturally lengthened the life of the chains, but too much would attract dirt, wearing the chain prematurely. Chain final drive was a less than ideal mechanism, and the gear drive solved the worst of the problems. There was nothing to adjust, the mechanism was fully enclosed and shielded from dirt, and power transmission was positive with little chance for slippage.

McCormick-Deering 15-30 Production	
Year	Production
1921	199
1922	1,350
1923	4,886
1924	7,321
1925	12,978
1926	20,001
1927	17,554
1928	35,525
1929	28,311
1930	21,891
1931	4,380
1932	1,705
Total Prod.	156,101

Source: McCormick/IHC Archives, 1921–32 "Tractor Production Schedule;" 1932–39 Serial numbers

The 15-30 was originally called the 12-25 Four-Cylinder International Tractor. The tractor was quickly changed to a 15-30 and geared down a bit. It seems the change in gearing resulted in the power upgrade, as the engine size and equipment were the same. In February of 1921, IHC ordered that a PTO be made optional on the 15-30. The PTO could be added to tractors already in the field, indicating that some 15-30s were built without a PTO. Vineyard editions of the tractor were built for California in 1922, as well.

Most interestingly, the decision to change the name from International 15-30 to McCormick-Deering 15-30 for the United States and Canada was not made until August of 1922. The model was introduced in 1921, making it likely that the first 15-30s were badged with "International" rather than "McCormick-Deering." Only about 200 were built in 1921, but all of them probably carried "International" badging. Also, the

The McCormick-Deering 15-30 first appeared in 1921, and was a new machine from the ground up. Designed to be built on a production line, it used an integral frame which was stronger and more dust-resistant than previous channel frames.

tractors exported off the continent continued to use the "International" name. The export tractors also used Deering and McCormick badging.

The engine was a vertical four-cylinder with removable cylinders. The crankshaft ran on ball bearings, an innovation that increased the main bearing's life span. The engine's bearing surfaces were fed oil by a combination splash and gear pump lubrication system. A high-tension magneto provided the spark, with a governor maintaining engine rpm at 1,000. The engine was started on gasoline and switched over to kerosene once it was warm.

The 5,000-pound tractor sat on an 85-inch wheelbase and could turn within a 30-foot radius. Standard equipment included a belt pulley, lugs, air cleaner, tools, and front wheel skid rings. Priced at around $1,250, the tractor was not cheap, but it was a good value because of its quality design and construction.

Specialty Models and Attachments

In 1923, rice field equipment was offered for the tractor, which consisted of widened wheels with special lugs and special front-wheel tire ring attachments. McCormick-Deering 15-30s were soon after offered as orchard models, which used lower wheels, a shortened air intake, and had the pulley and pulley carrier removed. The California orchard models, released in April of 1924, received the above modifications and rear wheel fenders that covered the top half of the rear wheel and swooped over to meet the hood.

Few tractors were produced in the early years of production, which may have been a result of some early production difficulties. By 1927, production was up to 17,000 a year.

In 1926, the McCormick-Deering's transmission was changed to lower the gear ratios for more drawbar pull. The bevel pinion and shaft, drive bevel gear, low-speed gear, medium- and low-speed pinions, medium-speed gear, and reverse pinion and bushing were replaced. The new transmission was not interchangeable with the old.

Beginning in January of 1928, rubber-skinned wheels were available for the McCormick-Deering 15-30. Such tires were strips of rubber wrapped around steel wheels, providing some relief to the problem of traversing pavement. Such tires were useful for both industrial and orchard applications. Some orchard growers found that steel lugs damaged roots and smooth wheels provided insufficient traction. These early rubber-coated wheels were less than satisfactory for field work. Until Firestone adapted the pneumatic tire to agricultural use in the early 1930s, lugged steel wheels remained the farmer's most efficient choice.

More Power for the 15-30

In August of 1928, the McCormick-Deering 15-30 was given major updates which resulted in a 22-36 power output. The factory continued to call the upgraded model the McCormick-Deering 15-30. Sometime in 1930, the sales department requested the name be changed to 22-36 to reflect the power upgrade. The model was badged as the McCormick-Deering 22-36 for a while, but the name did not stick and these higher-powered McCormick-Deerings are labeled 15-30s more often than not.

The additional power came from increasing the bore of the engine cylinders from 4 1/2 inches to 4 3/4 inches. The crankshaft and connecting rods were strengthened, and the cylinder head was new, with larger ports. The cooling system was still enclosed with a radiator, but a water pump was added. The intake and exhaust manifolds were new, and the carburetor was described as using water and fuel, probably some kind of crude water-injection system. The engine was governed to run 50 rpm higher, at 1,050 rpm. An oil filter was incorporated, and the

McCormick-Deering 10-20 Production

Year	Production
1923	7,117
1924	11,197
1925	18,436
1926	25,021
1927	26,646
1928	30,353
1929	39,433
1930	32,230
1931	10,901
1932	1,852
1933	20
1934	1,872
1935	1,334
1936	2,220
1937	2,301
1938	1,895
1939	742
Total Prod.	213,570

Serial Number Codes
Source: McCormick/IHC Archives, "Tractor Production Schedule" (1923–31); McCormick/IHC Archives, "10/20 McCormick-Deering Tractors-Date Built" (1932-1939)

10-20 Narrow Tread

Year	Production
1925	8
1926	148
1927	183
1928	323
1929	383
1930	207
1931	83
1932	79
1933	46
1934	3
Total Prod.	1,463

transmission and rear differential were strengthened. The strengthened parts increased the weight somewhat, as the tractor weighed 6,500 pounds with the standard lugs, pulley, tools, air cleaner, front wheel skid rings, and rear wheel fenders.

The upgraded model and the end of the Fordson in 1928 put some life into sales, pushing unit sales above 35,000 in 1928. In 1933, pneumatic tires became available through special order for the McCormick-Deering 15-30. Production of the tractor tapered off soon after, and the model was discontinued in 1934.

There were apparently some problems with the PTO on the 15-30 and 22-36s, as an upgraded PTO shaft and other parts were released in 1928, although the new part was not publicly announced. Also, some troubles with the rear end led to several new pieces and improved seals to protect against dirt penetration.

McCormick-Deering

10-20

Not long after the decision was made to go ahead with the McCormick-Deering 15-30 (or International 15-30, as it was known in 1921), IHC decided to build a smaller model. Like the

n integral frame and gear final drive. The same high-ten-
throttle-type governor, gas/kerosene fuel system, combi-
and-gear pump engine lubrication, and 1,000 rpm engine
speed were used. Cooling was the same; an enclosed system with a radiator but no water pump.

Where the new machine differed was simply in size. It used a smaller
bore and stroke at 4 1/4x5 inches. The tractor's dimensions were smaller
than the 15-30's in almost every aspect. The 78-inch wheelbase was 7
inches shorter, the 123-inch overall length was 10 inches shorter, and
the wheels were 8 inches smaller in the rear and 4 inches smaller in the
front. The 10-20 weighed about 3,700 pounds, compared to the 15-30's
5,000.

Early Production

While the official decision to build 10-20s was recorded in August of
1921, it seems none were actually built until 1923. Note that official or
company decisions are documented records of changes and new models.
Although company actions may lag behind company decisions, the official decisions are the best source for when new models were built.

Factory records show several related decisions after the August 1921
decision, all ordering that this new 10-20 be built. The original decision
is covered in pen marks, as someone apparently wanted the machine a
bit smaller than proposed. It also read, "Do not send copy of Decision to
Branch Managers until advised by the Sales Department to do so," indicating that the details were not yet satisfactory. The next decision men-
tioning the 10-20 shows up in September of 1922, over a year later. The
decision specifies that the 10-20, formerly known as the International
10-20, would be badged as the McCormick-Deering 10-20. The report
indicates the tractor had not yet been manufactured, supporting the belief that 10-20 production began in 1923. On December 22, 1922, another decision came out authorizing production of the "new" McCormick-
Deering machine. Again, the branch managers were not to be notified
until the sales department gave the green light. Apparently, they did,
because over 7,000 were produced in 1923, the 10-20's inaugural year.

Specialty Models

In 1926, the Narrow Tread McCormick-Deering 10-20 was approved
for production. Intended for use in orchards and other places that required tractors with a narrower tread, it had a 50-inch rear tread width,

The 10-20 Industrial became known as the Model 20. It was introduced in 1923 as a lightly modified McCormick-Deering 10-20, and was available until 1940.

which was about 10 inches narrower than that of the standard 10-20. The significance of the narrow tread machines was that altering tread width required new rear cases and significant costs for new tooling, changing production, promotion, and so on. Despite the cost, there were still only two different tread width 10-20s to suit the needs of a globe covered with farmers. This was a problem IHC would look to solve with the Farmall.

In 1933, the pneumatic tire became available on the McCormick-Deering 10-20 through special order.

Industrial Models

From the very beginning, IHC had a knack for getting the most from its tractors. Titan Road Rollers were a classic example of an adaptation of existing machinery to a new market. These machines provided a profitable sideline for IHC due to the fact that engineering and production costs were lower than building an entirely new machine.

But road rollers and graders were fairly complex conversions that would eventually be overshadowed by equipment designed specifically for such tasks. A converted tractor could not compete with a machine tailored to grade or roll roads, and the conversions were not especially cheap or efficient.

About the time the new McCormick-Deering machines were being introduced, IHC began to explore the potential for a simpler brand of specialty machine. Standard tread tractors were useful for far more than simply plowing; the company merely needed to get the tractors into the right location. Rather than complex conversions, a utility tractor could be created with little more than a new set of stickers and wheels.

Despite the simplicity of creating industrial machines, getting into the industrial market was an entirely different matter. With a concentrated effort of advertisements and sales people, IHC put its tractors to work on an incredibly diverse range of jobs.

Part of the company's effort was to get articles into newspapers and magazines touting the money industrial tractors saved cities and townships. The articles were written with a slant that is somewhat entertaining today; IHC didn't pull any punches with press coverage. This example comes from the *Chicago Daily Tribune* of June 25, 1924:

"It has been demonstrated in dollars and sense that skinny old nags and bony old plugs, some lame and some blind, are too expensive to carry on the city pay rolls, even if they have aided many a captain to carry his precinct in a primary or an election."

The article went on, lambasting politicians for dragging their feet and costing the city of Chicago as much as $700,000 annually. A series of articles appeared on the topic, and the third installment cited a savings of nearly $1.5 million by purchasing industrial tractors. The fact that to save $1.5 million, Chicago would have to place a $2 million order with IHC was downplayed, of course. While today's media are branded as untrustworthy, the subtle slant of modern coverage pales compared to the outright lies considered acceptable practice for newspapers of the 1920s.

However questionable the press of the day may have been, such articles were simply good business at the time. The tactics worked well, and industrial tractors became a long-standing success for IHC. These tractors were the grunts of the company. They showed up in unusual places, and certainly led more exciting lives than the Farmalls, most of which went straight to the farm. They were also a very profitable sideline for IHC, and the company made such tractors until the bitter end.

Model 20

The McCormick-Deering 10-20 was adapted for industrial use in 1922. The Model 20 serial numbers are indistinguishable from other McCormick-Deering 10-20s. Cast iron disc wheels were used front and rear, and only 23 of these machines were built in 1923. IHC realized that more was required than simply different wheels to make a truly useful industrial tractor. A foot accelerator, suspended front end, high-back seat, underslung muffler, dual rear tires, and rear-wheel brakes were added in 1924, with only modest sales increases.

Development continued for 1925, and a high-speed (10.4 miles per hour) top gear and transverse leaf front suspension were added. Front and rear bumpers and an assortment of wheel lugs were added as options. Sales for 1925 are listed as "several hundred" and IHC stated, "there was considerable market for this tractor in modified form." The company responded to the demand by adding an "IND" suffix to the serial number in 1925. In 1926, production of Model 20s reached 1,400 units, and the Model 20 was assigned unique serial numbers that began at 501 and were coded "IN."

Model 20 Production

Year	Production
1923	NA
1924	NA
1925	NA
1926	1,204
1927	1,842
1928	3,048
1929	4,607
1930	3,397
1931	1,831
1932	715
1933	206
1934	209
1935	173
1936	195
1937	212
1938	120
1939	127
1940	32
Total Prod.	17,918

Source: McCormick/IHC Archives

Engine and Specifications

The Model 20 used a four-cylinder valve-in-head IHC engine. Bore was 4 1/2 inches and stroke was 5 inches, and the engine was rated to produce 25 horsepower at 1,000 rpm. Removable cylinder sleeves were used, and the crankshaft spun on roller bearings. The engine and transmission were lubricated with a combination oil bath and oil pump system. An oil filter was used, as was an air filter. The engine was cooled by a thermosyphon system that used a radiator and a fan.

Later versions, at the least, used a four-speed transmission.

Top speed was about 10 miles per hour, with the three lower gears pulling 2.3, 4, and 7 miles per hour. The intake and muffler were mounted to the left side, and didn't protrude much higher than the hood, presumably for ease of use on the factory floor.

The tractor weighed in between 5,180 to 7,450 pounds, depending on the type of wheels used. Standard equipment included a cushion spring seat, hand and foot brakes, spring-cushioned drawbar, a muffler, and radiator curtains. Options included a belt pulley, 543-rpm power take-off, electric and gas lights, fenders, and a tire pump. In addition, IHC would custom-build the tractors. The following modifications were made in the late 1920s to industrial models: special equipment for a dump body type tractor, full reverse transmission, wheels with removable solid rear tires, Bendix Westinghouse air brake equipment, water muffler attachment, narrow tread tractor for sidewalk plows, and City of New York crosswalk plows

The Model 20 continued to be built through 1940. In fact, the first Nebraska Tractor Test of the machine took place in July of 1931. The machine put out 23.01 horsepower at the drawbar in low gear and 29.87 horsepower at the belt pulley. As tested, the tractor weighed 5,415 pounds and was equipped with an IHC air filter, governor, and E4A magneto. A Zenith C5FE carburetor was used.

Model 30

The modest success of the Model 20 was apparently enough to convince IHC that the industrial market was worth pursuing, and a prototype industrial version of McCormick-Deering 15-30 was constructed early in 1929. The "new" machine was dubbed the Model 30 and had the same general features as the Model 20. Production began in 1930, and very few of the machines were built.

An IHC industrial tractor brochure lists the Model 20, Model 30, and I-30. Looking closely at the specs, it appears that the Model 30 is indeed little more than a McCormick-Deering 15-30 with some special equipment. Company records indicate that the Model 30 serial numbers were simply 15-30 numbers stamped with the prefix "HD."

Model 30 Production	
Year	Production
1930	48
1931	326
1932	158
Total Prod.	532
Source: McCormick/IHC Archives	

Engine and Specifications

The Model 30 used a higher-powered version of the 4 3/4x6-inch four-cylinder engine used in later McCormick-Deering 15-30s. The Model 30 engine was rated for 45 horsepower and was governed to run at 1,050 rpm. The engine used the standard IHC removable sleeves, ball-bearing crank, and combination pressure/splash lubrication. A three-speed transmission was used, with a fairly low top speed of 5.9 miles per hour. The coolant system used a radiator with a fan, water pump, and thermostat.

The Model 30 weighed a solid 9,700 pounds, and swelled to 11,700 pounds when equipped with dual rear tires. The machine outweighed the original McCormick-Deering by over 4,000 pounds, probably due to a combination of heavy solid wheels and rugged front suspension.

Production ended in 1932, with only 532 Model 30s constructed. This extremely low number makes the Model 30 one of the rarest IHC tractors built, and a great find for the collector.

The Model 20 and 30 seeded the ground for a long line of industrial machines to come. With each new standard tread machine, an industrial variant would be produced. Custom fabrication continued in the industrial line, and these tractors appeared in more unusual and creative forms than any other type of IHC tractor.

The McCormick-Deering machines and their variants were well-built, reliable, and effective farm tractors. They did exactly what a tractor of the 1920s was expected to do better than anyone had reason to expect. But they did not break any new ground. They simply covered the open ground well.

To compete with Ford, IHC needed the McCormick-Deering machines. But to beat Ford, it needed something revolutionary, something no one had seen. Bert R. Benjamin had that tractor under his hat, and after a long battle, he would use it to finally put Ford out of domestic tractor production. But that, of course, is another chapter.

W-30

Standard tread tractors were still a priority in the early 1930s, and the McCormick-Deerings were in need of an update. The W-30 was just that; an upgrade of the McCormick-Deering 10-20. The increased-power engine was bumped up to about 30 horsepower, hence the designation.

The W-30's engine put out a little more horsepower than the Farmall

F-30's. The bore and stroke were the same at 4 1/4x5 inches, but the engine was rated for 31.63 horsepower at the belt, which was 1.3 horsepower more than the F-30. At the drawbar, the two were rated nearly equally. Both engines were governed to run at 1,150 rpm, although the W-30 was rated to run at 1,300 rpm late in production (which was done by using a new governor spring).

The W-30 engine used a combination manifold that allowed it to burn kerosene/distillate or gasoline. A Zenith carburetor was used, as was an International E4A magneto. A thermostat-regulated radiator and water pump cooled the engine. Power was transmitted to the transmission via a 12-inch single-plate clutch, and the brakes were hand-operated.

W-30 Production	
Year	Production
1932	11
1933	26
1934	2,634*
1935	6,541*
1936	6,236*
1937	7,875*
1938	6,088*
1939	2,560*
1940	560*
Total Prod.	32,531

*Number of serial numbers issued that year, which is only an approximation of production numbers. Source: McCormick/IHC Archives, 1932 and 1933 from "Tractor Production Schedule"

Although the engine was nearly identical to that of the F-30, the chassis was vastly different. The previously mentioned integral frame was used, but the front and rear axles and final drive were all unique to the wheeled 30s.

The transmission is a bit of an enigma, as early company records list a three-speed transmission. At some point in production, a four-speed transmission was adopted. In 1936, a lower-geared four-speed transmission was fit, and serial numbers were changed with the improvement. The decision to make the change was made early in 1936, but a few of the older units were sold into 1937.

Options for the W-30 included pneumatic tires, high-altitude pistons, a lower low gear, differential lock, and a distillate-burning manifold.

Less than 50 W-30s were built in the first two years of production. The tractor was finally sold in quantity in 1934, when 2,634 rolled out of the factory doors. Sales remained steady until 1940, when the W-30 was being phased out in preparation for the W-6.

The W-30 served IHC well. Although the company was able to use parts and technology designed for the McCormick-Deering 15-30, the cost of building a custom rear drive and front end made the W-30 expensive for the volume of sales produced.

Model I-30

When the W-30 appeared, it was quickly pressed into duty as an addition to the Industrial tractor line as the I-30. It fit neatly in between the Model 20 and Model 30, giving greater flexibility to the line.

The experience the company had with industrial tractors brought a number of improvements to the I-30. The drawbar was strengthened, and built to be more adaptable to mount equipment. Some I-30s were fitted with equipment to perform specialized tasks ranging from landing dirigibles to operating crosswalk plows. The I-30 was produced at Milwaukee Works.

The I-30's specifications closely resembled those of the W-30, which was the base model for the machine. The I-30 was slightly smaller than the Model 30, and weighed more than 2,500 pounds less with about 10 fewer horsepower.

The I-30 engine was an IHC valve-in-head four-cylinder with a 4 1/4 x 5-inch bore and stroke. The engine was rated for 35 horsepower and governed to run at 1,150 rpm. Like most of the IHC line, it used removable cylinder sleeves, ball-bearing crankshaft journals, and splash/pump lubrication. The cooling system used a radiator, water pump, and thermostat.

Options included a belt pulley, 575-rpm PTO, electric and gas lights, fenders, special wheel equipment, and custom-built equipment.

W-40

With the market's demand for horsepower steadily increasing, IHC responded with a powerful wheel tractor, the W-40. The new tractor used a heavy-duty integral frame mated to the six-cylinder engine from the TA-40 TracTracTor. Seven experimental versions of this rugged machine were built for testing in 1934.

The W-40 was first designated the as WA-40, then the WK-40, and finally simply the W-40. Each variant is identified by a unique serial number code, and

WA-40, WK-40, W-40, and WD-40 Production

Year	Production
1934	NA
1935	940*
1936	3,679*
1937	2,545*
1938	2,091*
1939	567*
1940	237*
Total Prod.	10,059

*Number of serial numbers issued that year, which is only an approximation of production numbers.
Source: McCormick/IHC Archives

some bore badging designating "WA-40" or "WK-40."

The six-cylinder 3 5/8 x 4 1/2-inch engine ran at 1,600 rpm and was rated for 45 horsepower at the belt and 28 horsepower on the drawbar. The engine was started on gasoline and switched over to kerosene or distillate. A 14-inch single-plate clutch transferred power to the rear end. The trans-mission was a three-speed, and the front and rear axles were similar to those used on the W-30. The tractor weighed in at 6,100 pounds.

I-40 and ID-40 Production	
Year	Production
1936	20*
1937	98*
1938	58*
1939	94*
1940	79*
Total Prod.	349

Number of serial numbers issued that year, which are only approximations of production numbers.
Source: McCormick/IHC Archives

WD-40

At the same time as the W-40 prototypes were authorized, the company decided to built another prototype with the same chassis and the four-cylinder diesel engine from the TD-40 crawler. The diesel engine produced about the same amount of horsepower as the six-cylinder kerosene-burning engine, although at a lower rpm of 1,100. The resultant tractor, the WD-40, was released in 1934, and was one of the first production diesel tractors. The diesel engine used a unique gasoline starting system that switched to diesel once the engine was warm.

Both tractors tested well, and were released in limited numbers for regular production in 1935. The transmissions on the prototypes were a bit fragile, and a strengthened unit was authorized for the regular production machines. Several gears and gear shafts as well as bearing cages and the transmission case were strengthened.

The heavy field tractors sold well for about three years, and died off in 1939 and 1940. At the end of their run, they were replaced by the W-9 series. The W-9 would be the only remaining wheel tractor that was independently designed rather than based on a cultivating tractor chassis.

I-40

The I-40 and ID-40 were developed in 1935 and production began at Milwaukee Works in 1936. The heavy-duty machines were fitted with the standard industrial equipment—front and rear bolsters, solid wheels, and any custom equipment desired by the purchaser. Purchasers, sadly, were in short supply, making the I-40 and diesel ID-40 rare today.

The W-40 was an upgraded version of the McCormick-Deering 15-30, with more horsepower in a similar package. The W-30 (and the WK-30) was the standard tread counterpart to the Farmall F-30, but the two shared little more than the "30" in their model names.

Tough Times

The McCormick-Deerings came onto the market in the early 1920s, at a time when the American economy was on the rise and the farm economy was struggling. Farmers continued to struggle until World War II, battling through cutbacks in demand after World War I and the Great Depression.

Tractor companies invested in agriculture throughout this period, and developed new machines at a surprisingly vigorous rate despite a market that was in turmoil. The company leaders appeared to understand that the market would improve and that mechanization was coming to the farm. It was vital, and the enticement was that the manufacturer who delivered the most appropriate products would get rich.

Henry Ford brought the price of a farm tractor down, and increased the expectations of the farmer. International's McCormick-Deering line combined high-quality componentry and technology with an evolutionary design. The tractors were amazingly well-built, and IHC honors the lifetime guarantee on the engine bearings yet today. Try to find another product for which the manufacturer will replace parts more than 80 years after it was produced.

The flaw of this line was that it didn't change the farmer's situation. The McCormick-Deerings and Fordsons were good machines, but they didn't completely eliminate the need to have horses to cultivate.

So farmers were still dealing with livestock, meaning that time on the field was limited by the endurance of your horse and tasks like cultivating were time-consuming.

The McCormick-Deerings were created by a team of engineers within IHC, and their design reflected that. In order to create the machine that would truly replace the horse, vision was required. International had that power in engineer Bert R. Benjamin and engineering team leader Alexander Legge. Benjamin would provide the vision and commitment to make the machine the agricultural world needed, and Legge was the advocate who kept Benjamin at work despite the skepticism of a giant corporation.

Chapter Three

One Man's Vision

Bert Benjamin and the Farmall

"For there came a time when but for the persistence
of B. R. Benjamin, the whole idea
might have been abandoned."
—C. W. Gray, 1932

While International and Ford battled to control the tractor market, IHC was waging civil war internally. The planned counterstrike to the Fordson was the McCormick-Deering line. In board rooms and thousands of internal memos, the strategy to compete had been outlined carefully. The result was what you might expect from a large company—a meticulously thought out, well-built machine that took no chances.

In another department, engineer Bert R. Benjamin was working hard to create a more innovative machine—the Farm-All. His wacky new designs were evolutions of the Motor Cultivator, a failed experiment, and his early prototypes looked odd and worked, well, like early prototypes. All this meant that his creations were not greeted by the company with welcoming arms. In fact, there were those who wanted to have Benjamin reassigned and his strange little project stuck back on the shelf.

With help from the farmers fortunate enough to test experimentals, the sympathetic ear of International executive Alexander Legge, the Farmall survived to become arguably the most significant tractor in the history of the International Harvester Company.

The McCormick-Deerings were very capable machines, but they were not especially revolutionary. They used state-of-the-art componentry and production techniques, but the basic design of these machines had been around for nearly a decade. The tractors did everything a tractor of the day was expected to do very well, but they didn't do anything unexpected.

The innovation demanded by the times and sought for by industry was the all-purpose tractor. Tractor farming was accepted as an improvement and tractor production soared, but the farmers of the day had yet to be convinced that the horse could be completely replaced on the farm. In fact, the number of horses on American farms increased until 1920, and did not drop off significantly during the 1920s.

International had an innovative new machine in the works that could cultivate (and therefore replace the horse), but that machine would have to overcome the corporate inertia backing the McCormick-Deering machines, not to mention some nasty comments about its spindly looks.

Seeds for the all-purpose tractor were planted before IHC formed. One of these early machines was Edward A. Johnston's Auto Mower, which he developed for Deering. Johnston was by all accounts a brilliant designer who engineered the International Auto Buggy, the vehicle that pioneered IHC's truck line, and his Auto Mower won a prize at the World's Fair in Paris in 1900 in head-to-head competition against the McCormick Auto Mower. Johnston understood the needs of the times, and he would continue to play a role as the Farmall developed.

John F. Steward was another forward-thinking IHC engineer who planted seeds for the Farmall. As early as 1902, he was working on a motor-driven harvesting machine. Eight years later, he proposed a multi-purpose tractor to IHC executive Alexander Legge. Stewart wrote, "What we want is a tractor that will most nearly abolish expensive labor on the farm, both of brute and of man."

Steward's writing was prophetic, but his relationship with the McCormick-run IHC was not a warm one. Steward was an ex-Deering Harvester Company employee who never forgot that the McCormicks and the Deerings were once bitter competitors.

Motor Cultivator

One of the first machines to result from the search for the general purpose tractor was the motor or power cultivator. Tractors of the time pulled a plow and turned a belt acceptably; what was missing was the ability to cultivate efficiently. The Motor Cultivator was developed to meet this need.

The machine was as simple as its name; it was a self-propelled cultivator. Carl Mott and Edward Johnston patented a design for a Motor Cultivator in 1916, and set about building experimentals. The quest to build a motor cultivator spread across the industry, as most of the major manufacturers were experimenting with some sort of vehicle designed simply for cultivation. In addition to the International model, Allis-Chalmers, Avery, Bailor, and several other lesser-known companies had motor cultivators on the market in the late 1910s.

The International Motor Cultivator was a four-wheel tricycle design, with the two rear drive wheels close together and mounted on a swivel. The machine was turned by pivoting the rear wheels. Out front were two tall, narrow wheels, spaced widely apart. The cultivator hung below the chassis and between the front wheels.

The machine was powered by a four-cylinder engine with a bore and stroke of 3 1/8 x 4 1/2 inches rated to produce 12 horsepower at 1,000 rpm. The motor was water-cooled with a radiator and fan that used thermosyphon circulation. The engine was mounted to a channel steel frame, and the entire unit weighed about 2,200 pounds.

The machine was designed specifically for corn cultivation. Early tests of the Motor Cultivator showed that it would overturn when driven on side hills. A nose weight partially solved this problem, but the machine still tipped over too easily. In 1917, the company authorized commercial production of 300 Motor Cultivators. Only 31 made it into the hands of farmers, and they were not well-received.

The Motor Cultivator fared poorly in the fields. It was slow and under-powered, and the wheels broke corn roots. Under ideal conditions, the machine performed adequately. Ideal conditions were relatively rare on the farm. The Motor Cultivator's questionable performance and short supply effectively killed it for 1917. By the time the factory assembled about 100 machines, it was so late in the season that the sales staff felt it could not sell the remaining stock of just over 50 machines.

In 1918, heavier cast rear wheels were added to the Motor Cultivator. The rear wheels were added to offset the machine's tendency to tip. The

The Farmall was engineer Bert R. Benjamin's vision, which he brought to fruition with the help of IHC president Alexander Legge. This version of the Farmall is a later experimental, with all the fundamental pieces in place. It was photographed at Ames,

wheels helped, but the design was inherently unstable. The engine was carried high on the chassis. When the rear wheels were castored to turn the machine, the rear placed a rotating force on the entire machine. If the Motor Cultivator was turned downhill, it would tip over.

To worsen matters, the 1918 model was again delivered late. Dealers were promised 150 models by June 10. They didn't show up until July 15. By the end of 1918, 301 Motor Cultivators had been built.

In August of 1918, the final nail was placed in the Motor Cultivator's coffin. Cost estimates for the 1919 model were high, at over $500 for a machine that would retail for about $650. Revised cost figures came in, and the retail price would have had to be over $800 for the company to make a profit. Avery's motor cultivator sold for $540, which meant that $800 was too high for the market to bear. On August 29th, the 1919 order for the Motor Cultivator was canceled.

In the report "Notes on the Development of the Farmall Tractor," C. W. Gray wrote:

"The rock on which the Motor Cultivator finally broke was manufacturing cost. One wonders now how much of that excessive cost was due to the weight which had been added in the effort to make a wrong principle work right. One wonders, indeed, whether the application of some very simple scientific formula should not have disclosed early in 1917 that the Motor Cultivator would probably roll over on side hills. One wonders, again, why, when it was discovered that the outfit would upset, someone did not foresee the weight which would be required to stabilize a tractor so designed must prove prohibitive. Yet we went right on, empirically adding weight until weight killed the Motor Cultivator. And there is good reason to believe that the abortive attempt to market the Motor Cultivator all but completely discouraged the management in its purpose to build an all-purpose tractor."

Quest for the General Purpose Tractor

The Motor Cultivator was discontinued. Slow, tippy, poorly designed, and over cost, it never really had much of a chance. But the engineering department—especially Benjamin—had not given up on the concept. Years later, Benjamin recalled his feelings on the matter:

"There was talk about a new kind of tractor in the industry, but no one had such a machine or even much of an idea on how to start building one. I knew we had to come up with an all-purpose tractor—one with rear wheels that could be adjusted to straddle two rows of crops and a narrow front with a wheel that would fit into one crop row."

In the late 1910s, Benjamin and his engineers first put these ideas on paper, and began work on an experimental model. Benjamin said, "Working with D. B. Baker, then chief engineer, and John Anthony, who took care of the drafting, we had an operating experimental model within three years."

So began the evolution of the Farmall. In 1920, the drive wheels were shifted from the narrow rear to the wide front. This move was probably prompted by Legge's insistence that the machine, "at least look like a tractor." This version also featured the automatic differential brake that was used on the Farmall. Note also that the "front" and "rear" of the tractor was a slippery concept; the 1920 version sported three forward and three reverse gears and a reversible seat so that it could run either way. Also, the engine was rotated and was mounted in-line rather than transversely.

This machine was known as the "cultivating tractor," and late 1920 photographs show versions of the machine bearing the name, "Farm-all." The tractor was reviewed in an IHC memorandum as too cumbersome, too heavy, the steering action slow, and too difficult to get sufficient gang shift.

The 1920 Farmall had some serious teething problems to solve, but Benjamin was still a firm believer in the basic design and concept. IHC executives were divided on the topic, and the upcoming release of the all-new McCormick-Deerings did not help Benjamin's cause.

Benjamin provided evidence of the Farmall's viability by comparing the International 8-16, the Moline Universal, and a proposed Farmall 8-16. Each machine was evaluated by it's ability to operate eleven commonly used pieces of machinery.

The Universal operated nine of the machines alone. The International 8-16 performed four operations without horses or additional persons. The Farmall could operate all 11 implements with only one operator. This was the true strength of the Farmall, and Benjamin had to play to it strongly to keep the project alive.

A film of a Farmall operating a 10-foot shocker and binder at Harvester Farm was shown to IHC executives on Dec. 13, 1920. Immediately afterward, the group met to discuss the fate of the Farmall program for 1921. Most were unenthusiastic about the tractor. Despite the tractor's potential as an all-purpose machine, it was pointed out that the tractor and all of the accompanying implements would be a sizable investment and that using the array of attachments would be complex for the average farmer. Also, the design was too heavy. Naturally, the failure of the Motor Cultivator was brought up as another negative aspect of the project.

Despite the fact that most IHC executives had little faith in the project, it was determined that further exploration of Benjamin's concept for a general purpose tractor was worthwhile. Five experimentals were authorized to be built in 1921, at a cost of over $150,000 because each example would be built by hand. Benjamin and a handful of engineers were reportedly the only people in the company with much enthusiasm left for the Farmall.

In January of 1921, the authorization for five tractors was changed to two lightweight versions of the Farmall. The Farmall of 1920 weighed about 4,000 pounds. The new prototypes were lighter.

At the same time, the cultivator was changed to mount to the front of the tractor. According to Arnold Johnson, an engineer working on Farmall attachments, the switch to a front-mounted cultivator went over well with his department and gained favor for the Farmall concept.

Benjamin kept a steady stream of information flowing to Legge. A document was prepared comparing costs between a farm using eight

The original Farmall became known as the Farmall Regular. Although the tricycle design looks natural enough today, it was quite unusual when compared to the standard tread tractors of its day. Alexander Legge, the company president, insisted that the Farmall be designed so it "at least looked like a tractor." A Texas cotton grower said, "It's homely as the Devil, but if you don't want to buy one you'd better stay off the seat."

horses, six horses and a Fordson, and a Farmall and two horses. The figures impressed Legge, and he passed the document around the company.

Benjamin then proposed a version of the Fordson tractor with the front wheels close together and raising the ground clearance in the rear. Such a machine could plow adequately and also cultivate. Benjamin felt such a machine would surely reach the market, and that it was only a matter of whether IHC or Ford built it.

Farmall Under Fire

In the meantime, the company called a special conference to discuss what could be done about Ford. The Fordson was killing IHC in sales, and solutions were hard to find. The Farmall program was discussed as a possible solution, and Alexander Legge sharply criticized the tractor's development. Johnston jumped to the Farmall's defense, and a chorus of lukewarm support followed. Some felt the Farmall could developed into a better machine than the Fordson, and most thought there was promise in the idea. The result was that 100 Farmalls were ordered to be built for 1922. Also, a new line of implements were to be developed for the McCormick-Deering machines.

The Farmall design continued to evolve, and the order to build 100 was reduced substantially to explore the new designs. The gangs on the cultivator were designed so that they shifted when the front wheel turned, allowing the operator to easily follow the rows when cultivating.

Sometime in 1921, the chain drive was abandoned in favor of an enclosed gear drive similar to the type used on the McCormick-Deering tractors. By the close of 1921, demonstrations of prototype Farmalls had gained the tractor some grudging respect but not much interest from the sales department. Legge stated development was too slow.

Although Legge made scathing comments in a memo about the Farmall, he also made sure Benjamin received enough support. Times were tough for the general purpose tractor program in 1921, and Legge knew that Benjamin was feeling down. Legge told a fellow employee, "Charlie, for heaven's sake go to Benjamin and see if you can give him some help and encouragement. He's got a real job on his hands and we want to see him succeed." Despite Legge's constant criticism of the Farmall project, he believed in the concept and in Benjamin.

The Farmall is Born

In January of 1922, the company authorized 20 machines to be built (company decisions are dated company documents ordering engineering actions such as new models, experimentals, and changes). The engine used a 3 3/4 x 5-inch bore and stroke and ran at 1,200 rpm. A throttle governor and high-tension magneto were used, and cooling was a closed thermosyphon system with a radiator and fan. Lubrication was splash complemented by a gear circulating pump, and the engine was started on gasoline and then switched over to kerosene. The tractor weighed approximately 3,500 pounds. A PTO and power lift were listed as standard equipment. This machine had a reversible seat and could be operated in either direction. Company records list 17 examples built.

At the time this decision was made, Legge was still skeptical of the Farmall. In a memo to a fellow executive, Legge wrote, "About the only thing we have settled so far is that we have done a very poor job of putting them [Farmalls] together, which suggests that you should try to strengthen your engineering staff."

The day after Legge's memo was written, Benjamin demonstrated a lighter version of the Farmall at the Hinsdale Farm. The machine apparently fared well. Before Legge left Hinsdale, the original order for 20 Farmalls was canceled, and the lighter version Benjamin demonstrated was to be developed further.

The Weight Debate

Early in 1923, the Farmall debate centered on power, weight, and cost. It was felt that the tractor needed to be capable of pulling two 12-inch plow bottoms and needed to weigh about 3,000 pounds. Cutting back features and quality to reduce cost was seen as a mistake. Note that the weight of the Farmall would creep to just over 4,000 pounds by 1925.

The manufacturing arm of IHC was occupied with building McCormick-Deering tractors, so the engineering department would have to construct the 1923 machines. Also, it was decided that the tractors needed to be capable of pulling existing implements as well as custom-made units. Beyond the production issues, some feared that the release of the Farmall would harm McCormick-Deering 10-20 sales. A maximum of 25 Farmalls were authorized to be built in 1923, along with 25 cultivating attachments. The specifications for the machines were essentially the same as those of the 1922 machine, with the exception of trimming about 300 pounds of weight.

At the same time, Legge was discouraged to discover that the engineering department was developing several types of rear gear drive. He felt the tractor had been developed enough, and that there was not time nor resources for drastic engineering changes. Legge wrote, "I think perhaps it is true that in the past we have brought out some of our designs a little too hastily; however, this is not true of the Farmall, as you have been building it up and down for five or six years and by this time ought to have a pretty fair judgment as to what is the best and cheapest plan to follow." Whatever the new development may have been, Legge's letter put an end to it.

The 25 prototypes were assembled and sent into the field. Performance reports were favorable. The tractor cultivated efficiently, steered easily, plowed well, was durable, and had respectable power on the belt.

The tractor was strengthened in an assortment of key areas for 1924. The bolster became a single casting, the bull gear housing was strengthened, and the rear axle was increased in diameter. The drawbar fastening was enlarged. The rear axle roller bearings were equipped with an oiler. Also, fuel capacity was increased to 13 gallons. The new tractor sold for a little less than a $1,000.

The production run was limited to 200 units. The numbers were kept small mainly to avoid hurting sales of the similarly-priced McCormick-Deering 10-20. The tractors were sold through dealerships, and each model was closely monitored to determine performance and reliability.

In the showroom, the tractor had some other hurdles to overcome. The tricycle design was unusual at the time, and the Farmall was rather gangly and unfinished when compared to the compact, more graceful lines of the popular standard tractors like the McCormick-Deerings. One Texas cotton grower said, "It's homely as the Devil, but if you don't want to buy one you'd better stay off the seat."

The tractor sold for $950 in 1924, and 200 were built and sold. Throughout 1924, the tractor's performance was lauded. Company executives believed the tractor would probably kill the McCormick-Deering 10-20, but also felt that it would be better for an IHC product (rather than a competitor's) to knock off the 10-20.

Strong Support From Texas

For 1925, the Farmall was again improved. The changes were fairly slight, but covered most of the tractor. A host of systems were simplified, including the gear shift, starting crank, steering gear, front bolster assembly, differential assembly, transmission assembly, and even the tool box. In addition, the wheel hubs and transmission case were strengthened and a muffler and air cleaner were added. The tractor weighed just over 4,000 pounds. International management's reluctance to put the Farmall into production resulted in a better machine. With each delay, the tractor was refined and improved.

Although only 250 were originally authorized to be built in 1925, the demand prompted IHC to build 838 Farmalls that year. The majority of these were sent to Texas, and were sold for $100 more than the 10-20. It was felt that the tractors would sell to cotton growers, and open the market to IHC without hurting 10-20 sales. Interestingly, IHC did not devote an advertising budget to the Farmall. This move reflected the company's desire to avoid interfering with McCormick-Deering sales. The publications of the time reflect the decision, as Farmall ads are scarce and IHC's magazine hardly mentioned the Farmall until 1926.

The reports were again favorable on the 1925 machines. In a company memo, G. A. Newgent responded to queries about how the Farmall was performing in Texas. Rather than reply in his own words, he gathered a few statements from farmers who were using the new machine. They included the following:

"I have used my Farmall eight months and on one occasion it was run six days and six nights without a stop except to refill it with fuel. I have tested it thoroughly and I am satisfied as the purchase of another recently proves."

• • •

"Well, I am buying my second Farmall today. When I bought the first last spring I was afraid it would give me considerable trouble as it looked spiderly, but there has not been a cent spent on it and it pulls as well today as it did the day it was delivered to me."

• • •

"My wife uses a nickel's worth of ice per day and the drippings from the ice box are more than ample to water my Farmall."

Newgent went on to write that the 250 Farmall owners were, without exception, extremely pleased with the Farmalls. He closed with this: "When a farmer buys a Farmall he joins the sales force of the International Harvester Company."

That was strong testimony, but another Texas sales rep had stronger testimony to offer. Several general office men—including representatives from sales, manufacturing, and engineer Bert Benjamin—met with a group of International's Texas salesmen. The Texas representatives wanted the company to build the Farmall for regular production. P. Y. Timmons, the general office manager of tractor sales, said that it might affect sales of the McCormick-Deering 10-20. The response was that the Texas branches didn't sell many 10-20s. J. F. Jones, general office sales manager, commented that if anyone was going to build a tractor that would displace the 10-20, it might as well be Harvester. Lastly, Jim Ryan, the manager at Houston, said, "If you don't adopt it for production, we will organize a company in Houston and build it down here."

This kind of statement was hard to ignore. Even so, there was still some concern about producing a tractor that competed with the McCormick-Deering. In December of 1925, Legge wrote that although the Farmall production for 1926 would be increased, he felt sales should be confined as much as possible to territories where its current lineup did not sell or work particularly well. Legge's concern was that the sales staff would put all its efforts toward the new Farmall and let down on promoting the other tractors.

Bert R. Benjamin

The Genius at McCormick Works

The design of the Farmall is largely credited to one IHC engineer, Bert R. Benjamin, whose persistence resulted in a revolutionary tractor that changed the industry approach to tractor design.

Benjamin grew up on a farm near Newton, Iowa, and studied agricultural engineering at the college in Ames now known as Iowa State University. He began his employment in 1893 with the McCormick Harvesting Company as a draftsman in the experimental department. In 1901, he became Chief Inspector of the McCormick Company, a post he held until 1902, when the International Harvester Company was formed. He continued in a similar position until 1910, when he was named Superintendent of the Experimental Department.

In 1922, Benjamin was named Assistant to the Chief Engineer at the IHC Corporate Offices, which was where he took the Farmall under his wing. He guided the Farmall through a series of set-backs and suffered criticism from those who could pull his funding, and triumphed. The Farmall proved all the nay-sayers wrong, and was perhaps Benjamin's defining moment.

Benjamin may have had a hand in the other farm tractor success story of first part of the century, as well. He was sent to the Ford plant to cooperate on work on the Fordson in 1917. A war board decision designed to increase production put him there.

Benjamin was granted 140 patents in his 47-year career with IHC. His work with the power take-off was another industry first, and the rest of his patents include harvesting equipment as well as a wide variety of tractor advances ranging from shifting cultivator gangs and the drawbar to the seat spring and draft couplings.

Columnist Elmer Baker wrote about Benjamin in a 1959 issue of *Implement and Tractor.* He had this to say:

"Then one man, not a committee, got an idea that a tractor that wouldn't displace the horses and mules on a farm was only half of the answer to the farm tractor. So he started to work on units that would... cultivate... as well as doing the land locomotive work of other tractors.

"That man was Bert R. Benjamin, and he worked for the Harvester company experimental department. He may have had an office on the old 7th floor at 606 S. Michigan Ave., Chicago, but we always found him somewhere around that old shack close to McCormick Works next to the Illinois and Michigan Ship Canal."

In 1937, Benjamin left the corporate offices and was assigned to research work, which he continued until his retirement in 1939. Benjamin was all engineer, as he remained in hands-on positions to the end of his career.

The American Society of Agricultural Engineers honored him with the 1943 Cyrus Hall McCormick gold medal. In 1968, Iowa State University's College of Engineering awarded him the Professional Achievement Citation in Engineering.

Benjamin died in October of 1969 at age 99.

Production Begins

In 1926, the Farmall's sales began to take off, and IHC was finally backing the Farmall concept. A new plant was opened specifically to build Farmall tractors. The plant was known as Farmall Works and, once it got started, tractors rolled out the doors at an incredible rate.

The machine received a generous splash of attention in *Tractor Farming*, IHC's promotional magazine. The Farmall was on the cover of the March-April 1926 issue, and the magazine lauded the merits of the company's "new" machine.

Farmall Regular and Fairway Production	
Year	Production
1924	200
1925	838
1926	4,430
1927	9,502
1928	24,899
1929	35,517
1930	42,093
1931	14,088
1932	3,080
Total Prod.	134,647

Note: Several IHC archive documents indicate that Regulars were sold in 1933 and 1934, although exact figures were not available; it is unclear (and doubtful) if tractors were still being built after 1932.

Source: McCormick/IHC Archives, "Tractor Production Schedule"

"For the last three years, this new tractor has been proving its mettle and merit in considerable numbers in a few limited sections, primarily the Southwest cotton territory.

"Quantity production is now getting under way, though it will be some time even yet before production can overtake the demand, especially now that the Farmall will be supplied, on order, to any section of the country."

The 1926 machine was improved again, and it certainly benefited from the long development phase. Each year, a few new problem areas were identified and strengthened. The result was that, by 1926, the Farmall was a highly refined machine.

Sales exploded to nearly 25,000 in 1928, and the Farmall was off to the races. It did just as IHC management hoped and feared. The McCormick-Deering 10-20 was pushed aside, as was the rest of the industry.

Narrow-Tread Farmall

A narrow-tread version of the Farmall was built to accommodate farmers who required a narrower wheel set. According to IHC documentation, the need for such a tractor was especially great in Argentina, so these tractors are probably quite rare in America. The tractor was equipped with a offset rear hub that narrowed the rear tread width from 80 inches to just over 63 inches. The narrow tread was authorized in the spring of 1927. Interestingly, IHC documentation also refers to a 57-inch rear-tread-width Farmall, which may have been a domestic narrow-tread version.

Another special edition of the Farmall was the Farmall Fairway, which was designed for use on golf courses. Smooth wheels were used, as well as offset wheels that narrowed the tread width to 57 inches. In 1929, a set of lugs was authorized for the golf course machines, probably due to wheel-spin on wet grass or hillsides. The lugs were simply small spikes.

Perhaps the rarest of the Farmall attachment packages was the orchard fenders. These were available from 1931 to 1934, and then were dropped due to lack of demand. This is easy to believe, as the ideal orchard tractor is low-slung, which the Farmall is not.

In 1930, a field change was supplied for early 1925 model Farmalls (serial numbers QC700 to 8821). The change package improved the tractors' durability, especially in dusty conditions. Most of the changes centered around better sealing against dirt. The change was authorized in February of 1930.

Farmalls Sold in 1933 and 1934

According to most documentation, production of the original Farmall ended in 1932. But an assortment of company documents indicate that a limited number of Farmalls were sold in 1933 and 1934. The most convincing of these is the record of payments to the Ronning family. The Ronnings owned a patent to several features similar to those of the Farmall, and IHC had to pay the Ronning family one dollar for every Farmall Regular, F-30, F-20, F-12, and F-14 sold.

Records of the payments for 1933 and 1934 show several hundred dollars paid for Farmall Regulars produced and sold. It seems highly unlikely that the company would make these payments unless the tractors had been sold. More likely than not, a few tractors from the 1932 production run were left over and sold in 1933 and 1934.

Rubber Tires

Near the end of the Farmall's production run, IHC authorized optional pneumatic tires for its tractor lineup. Although the decision wasn't signed until 1933, after the Regular was out of production, it listed 9.00 x 36 rear and 6.00 x 16 front pneumatic tires as additions for the Farmall. It is doubtful that many Farmall Regulars were equipped with these tires from the factory, but dealers and owners almost certainly retrofit pneumatic tires. The attachment number for the rear tires for the Regular was 25711-D; the number for the front tires was 25494-D.

A later retrofit was a hydraulic lift, which was authorized in 1945. The lift was a version of the new Lift-All adapted to Farmall Regular, F-20, and F-30 tractors. The part numbers were 350420R91 for the Regular, 350421R91 for the F-20, and 350422R91 for the F-30.

The Farmall design proved to have lasting merit, and the Farmall formed the basis of the IHC lineup for the next several decades. Engineer Benjamin succeeded in his quest and created an all-purpose tractor. The road to his success was filled with potholes and road blocks.

C. W. Gray claims the resistance to the Farmall was as much political as it was practical. Benjamin was often the only proponent for the Farmall, and Gray states that Benjamin threatened to leave unless the project received the funding and support he felt it deserved.

History owes a tip of the hat to Benjamin and the Farmall, as both brought in a new era of farm power. The horse, once vital to farming, was gradually replaced with the tractor, and the Farmall was the first tractor designed around and able to fill the horse's shoes. Some debated the merits of replacing flesh with steel, a companion animal with a cold steel beast, but it was impossible to debate the merits of increased productivity and decreased manual labor.

The general-purpose tractor had arrived. It appeared in the form of the Farmall.

Chapter Four

Tracked Solutions

The First TracTracTors

"The great thing in the world is not so much where we stand, as in what direction we are moving."
—Oliver Wendell Holmes

From the beginning, IHC was a diverse company. Beyond agricultural equipment, the company tried everything from cream separators and trucks to refridgerators and stoves.

One of the more logical product groups for the company to produce was crawlers. Companies like Holt, Best, and Cletrac demonstrated a definite demand for crawlers in orchards, the giant wheat fields of the Northwest, for logging operations, and on construction sites. The market potential was quite large, and IHC naturally looked to adapt its existing technology to crawlers.

With Henry Ford out of the way and Farmalls flying out of the showrooms as fast as IHC could build them, the company explored new markets and looked to expand in the late 1920s. Despite an oncoming economic slump, the scare with Fordson was a graphic demonstration of the fact that a large company cannot hang its hat on one or two product areas. For IHC to survive, it had to find success in some alternate markets.

The company experimented with four- and six-wheel designs of the International 8-16, the company's favorite test bed. The prototype tractors were quite popular with those lucky enough to get their hands on them. The four-and six-wheel-drive models were never put into regular production, but they provided some valuable engineering data on how to steer a vehicle without turning the front wheels.

The company first experimented seriously with crawlers during the late 1920s. The primary problem with adapting a wheel tractor chassis to crawler tracks was steering. The existing power train could use the rear axles to drive the track if the machine had only to go straight. Turning required separate activation of the left and right track. Stop the left, drive the right, and the tractor would pivot left. Reverse one and drive the other forward, and the tractor would spin in a circle.

To accomplish this, steering clutches were required to drive the left and right treads separately. Ideally, the clutches needed to be progressive as well. While sudden engagement was great for turning tightly or spinning around in corners, gradual turns would require simply letting off on one side or the other rather than total disengagement. To further complicate matters, track brakes were necessary to pivot neatly and to stop the machine.

So, the first challenge was to build a reliable, durable, and cost-efficient method to drive the tracks. When IHC got serious about building a crawler, it turned to machines the company knew well: the McCormick-Deering 10-20 and 15-30.

10-20 TracTracTor

The first IHC production crawler began with the McCormick-Deering 10-20. Perhaps the company speculated that a crawler might be a way to salvage the McCormick-Deering chassis after the inevitable loss of sales to the Farmall Regular. Perhaps the McCormick-Deering was simply well suited to conversion.

Initial experiments were with crawlers used a French & Hecht track adapted to an existing tractor. French & Hecht built track conversion kits for existing tractors, and gave IHC an existing mechanism to try out.

The company documentation of the 10-20 TracTracTor, or No. 20 as it later dubbed, is sketchy and contradictory. One 1940s document entitled "International Harvester Industrial Power Activities," traces the development of the tractors with recollections of engineers. According to that

TracTracTor 10-20 Production

(Also known as No. 20)

Year	Production
1928	NA
1929	472
1930	78
1931	975
Total Prod.	1,525

Source: McCormick/IHC Archives, "Tractor Production Schedule"

document, IHC began adapting the 10-20 to tracks in 1927.

Another document listing production figures shows that eight of the 10-20 TracTracTors were built in 1925, with a few hundred per year constructed until 1929 and less than a hundred per year in 1930 to 1932.

Total production figures are also contradictory, with *Endless Tracks in the Woods* listing 7,500 10-20 TracTracTors built and simple serial number math indicating about 1,500 were actually made. One company document says that 1,000 were built, while the company production chart from the archives lists 1,454 produced, which is consistent with the serial numbers. The 1924 date seems plausible enough, as the company was well on its way to recovering from the tractor wars at that point.

The 10-20 TracTracTors were built at Tractor Works. Problems with the early TracTracTors centered on track roller and bearing wear, mainly because they ran down in the dirt. Steel guards were built to cover the lower rollers, but the ultimate solution would be sealed bearings.

The 10-20 TracTracTor was tested at Nebraska in October 1931. It weighed 7,010 pounds, and was rated at 18.33 drawbar horsepower and 26.59 belt horsepower. The carburetor was a Zenith K5.

The crawler equipment made the TracTracTor about 2,000 pounds heavier than the McCormick-Deering 10-20. The additional weight and the traction afforded by the treads let the TracTracTor pull about twice as hard at the drawbar as the McCormick-Deering. The TracTracTor compared favorably to its crawler competition, as well. It was about 1,300 pounds heavier than the comparable Cletrac, the Model 15. The IHC machine out-pulled the Cletrac by about 20 percent. The TracTracTor also out-pulled the Caterpillar Model 20 at the drawbar, although the Cat pulled at a slightly higher speed.

Despite the TracTracTor's respectable showing, it was never produced in quantity. The machine convinces IHC that crawlers were a viable product, and the company continued to develop new models.

15-30 TracTracTor

The 15-30 TracTracTor was the crawler adaptation of the McCormick-Deering 15-30 wheel tractor. It seemed like a natural for an IHC crawler experiment, seeing as the company was already using the 10-20 as a crawler. Records on these machines are pretty scarce, but according to existing records, experimentation began in 1926 with an adaptation of the McCormick-Deering 15-30 that used brakes at the drive sprockets for differential steering. In 1928, the steering clutches from the 10-20 TracTracTor were adapted to the 15-30-based tractor. Work continued through 1929, and in 1930 the tractor was redesigned with steering clutches and brakes that could be serviced from the rear. You'll remember that steering clutches were the key to the crawlers, and that they were heavily abused and in need of regular adjustment and replacement.

In the document, "International Harvester Industrial Power Activities," this 1930 development work is discussed, as well as the new ability to service the brakes and clutches from the rear. The document also states specifically that the 15-30 TracTracTor was never put into production.

Model 15 TracTracTor

The Model 15 is the most intriguing of the early crawler models. A few photos are shown in *150 Years of International Harvester*, with little information. The machine is distinctly different than the 15-30 and 10-20 TracTracTor, with equal-sized track wheels, and a distinct guard over the pins and rollers.

According to the scant bits of information available, the Model 15 is the missing link between the 15-30 TracTracTor and the T-20. More interestingly, it appears that 50 Model 15s were actually built.

The IHC document, "International Harvester Industrial Power Activities," includes the following: "The smaller model which we then called the Model 15 TracTracTor was completed early in this year [1930] and went into production in 1931 when 50 machines were built. In 1932, full production was started and the name was changed to T-20 TracTracTor."

The "smaller model" is referenced in the section on the 15-30 TracTracTor as a smaller example of the redesigned 15-30. The new design is referred to as the forerunner of the new line of crawlers.

The final piece of the puzzle is found on the factory production sheet, which lists 50 T-20 TracTracTors built in 1931. The two numbers match,

The T-20 was the first of the TracTracTors to be produced in volume, and were popular for pulling combines in open grain fields and for logging operations. Its predecessors were a variety of crawlers adapted from the McCormick-Deering 10-20 and 15-30. The first 50 T-20s, produced in 1931 were badged as Model 15s.

which seems to indicate that the first statement—that 50 Model 15s were built in 1931—is true. Whether they were built as Model 15s or T-20s is the question, but if they truly exist as Model 15s, perhaps one has or will turn up somewhere.

T-20 TracTracTor

As stated above, the T-20 moniker was applied to the Model 15 in 1932. The T-20 was a bit more powerful than the Model 20, and used the F-20 power plant. The little TracTracTor was small and maneuverable, and was used in a wide variety of situations ranging from orchards to logging to grain harvesting. As you would expect, the crawler was more expensive than the wheel tractors. With a price of about $1,500 in

TracTracTor T-20 Production	
Year	Production
1931	50*
1932	1,502
1933	475
1934	746
1935	2,108
1936	3,114
1937	4,018
1938	2,036
1939	1,033
Total Prod.	15,082

*1931 machines produced as "Model 15"
Source: 1931–33 from McCormick/IHC Archives, Production Number Chart; 1934–on from 150 Years of International Harvester by C. H. Wendel

1935, it was about $400 more than its wheeled counterpart, the F-20. Despite the high price, the T-20 was fairly popular. About 15,000 units were sold from 1932 to 1939, when the new line of streamlined crawlers choked off the old T Series.

The T-20's four-cylinder IHC engine—which was the same as used in the F-20—had a bore of 3 3/4 inches and a stroke of 5 inches. The engine was rated at 31 horsepower and was governed to run at 1,250 rpm, 50 rpm higher than the F-20. An air cleaner, oil filter, and fuel filter were used, as was a three-speed transmission. The engine was started on gasoline and switched over to kerosene when warm.

The T-20 weighed 6,250 pounds, and came equipped with a 543-rpm power take-off. The T-20 put out about 25 horsepower at the drawbar. Standard tread width was 41 1/2 inches, although a 51-inch wide tread version was available as well.

The little crawler was quite maneuverable, and could be turned in a six-foot radius. A cab, canopy, belt pulley, front pull hook, and other options could be purchased for the T-20 on special order.

T-40 and TD-40

While small crawlers were great for jobs around the farm, the construction and logging markets had a need for large, powerful crawlers. For IHC to be player in these areas, it needed a big machine, and the T-40 and TD-40 were the entrants. They were developed early in 1931, and the T-40 used a six-cylinder engine. The TD-40 used a four-cylinder diesel engine in the same chassis. Both were built at Tractor Works.

In 1933, the gasoline version became known as the TA-40. The diesel TD-40 was introduced. The six-cylinder engines built in 1937 and after were equipped to burn kerosene or distillate. Note that advertisements sometimes listed the tractor simply as the T-40.

The TA- and TD-40 were physically larger and much more powerful than the T-20, with the gas tractor rated at 51 horsepower, with 44 horsepower available at the drawbar. The tractor was more than two tons heavier than the T-20, with the gas model weighing 11,200 pounds and the diesel an even 12,000 pounds.

A five-speed transmission was used, allowing the operator to exactly match power to load. Top speed was a galloping four miles per hour. The six-cylinder engine's bore was 3 5/8 inches and the stroke was 4 1/2 inches. The engine was governed to run at 1,600 rpm. A radiator and fan regulated by a thermostat cooled the big six.

TracTracTor T-40 and TD-40 Production

Year	Production
1932	232*
1933	476*
1934	850*
1935	1,608*
1936	1,910*
1937	1,416*
1938	880*
1939	967*
Total Prod.	7,631

Source: *from McCormick/IHC Archives serial number listings, so numbers are approximate; total is from 150 Years of International Harvester by C.H. Wendel

TracTracTor T-35 and TD-35 Production

Year	Production
1937	2,464
1938	1,659
1939	1,464
Total Prod.	5,587

Source: 1936 from 150 Years of International Harvester by C. H. Wendel; others from serial number data and are approximate

The tractor was available in a standard and wide-tread version, and sold for nearly $3,000, making it an expensive vehicle at the time, although not out of line for large crawlers. The T-40 and TD-40 could be equipped with an assortment of Bucyrus-Erie blades, transforming the crawlers into bulldozers. Options included a power take-off, enclosed cab, electric lights, various sizes of track shoes, and a canopy top.

T-35 and TD-35

The T-35 and TD-35 were slightly less-powerful versions of the big 40s. They were introduced in 1937, and were only produced for four years before the styled TracTracTors appeared in 1938.

The T-35 used a six-cylinder engine that could be equipped to burn distillate, kerosene, or gasoline. Burning gas, the big six put out 37 horsepower in Test G at Nebraska, while the TA-40 put out 44 horsepower in the same test, also burning gas. The six-cylinder had a 3 5/8 x 4 1/2-inch bore and stroke, and ran at 1,750 rpm.

The TD-35 used a diesel engine that produced 42 horsepower on the belt at 1,100 rpm. At the drawbar, the TD-35 put out about 37 horsepower. The TD-35 put out about 9 fewer horsepower than the big TD-40 on the belt, and weighed a few thousand pounds less as well.

Big Phantoms

Several large experimentals were at least tested in the mid-1930s, when the company experimented with a large crawler. At least five machines were mentioned in records: the T-80, T-65, TD-65, T-60, and TD-60. The development is discussed in "International Harvester Industrial Power Activities" as follows:

This TD-35 cutaway shows the five-speed transmission, diesel engine, and steering clutches. The T-35 used a combination of steering brakes and the steering clutches to turn.

"Development of this larger tractor [the TD-18] started early in 1934, when the Engineering Department built the T-80 which had the same rear-mounted steering arrangement as the T-20 and T-40 models. During 1935, we developed and built the T-65 with many improvements but in general along the same lines as the T-80. Then in 1936 we brought out the T- and TD-60 in which the steering clutches were essentially of the current production. In 1937 and 1938 this design was further developed and streamlined and was then known as the TD-65."

In *150 Years of International Harvester*, a drawing—not a photograph—is shown of a TD-65. The crawler is obviously styled. Although its difficult to tell without a size reference in the drawing, the TD-65 gives the impression of being quite large, with the platform and seat high off the ground and relatively small on the big machine. Author Wendel refers to a listing without production figures in IHC documentation, which is probably different than the text above. These few scraps of information give a tantalizing look at the development of IHC's large crawler, the TD-18, which was introduced later with the other styled crawlers. These machines may have been built experimentally and badged as TD-65s, but whether or not any of these machines survive today is unknown.

Chapter Five

Evolution of the Fittest

The Farmall Family Grows

"[The] tractor does two things—it turns the land and turns us off the land. There is
little difference between this tractor and a tank."
—John Steinbeck, *The Grapes of Wrath*

The power of mechanization collided with a disastrous economy in the 1930s. The tractor had not helped this situation terribly. The machines were expensive purchases for a farmer, and the tractor manufacturers made those sales happen by offering credit. When the debt load of America's growing lower class became too high, the end result was a market crash that left American farmers defaulting on loans and struggling to stay on their family lands.

The onslaught of power farming could not be slowed by a mere economic catastrophe. Tractor farming was more economical and efficient than horse farming, and the use of the horse declined steadily as more and more farms turned to tractors.

The manufacturers seemed to deal with the Depression with considerable aplomb. Production was scaled back, and profits slid significantly, but a surprising amount of research and development continued. Sales were strong for the tractor industry in 1930, at 200,000 units, but dropped off to only 19,000 tractors sold in 1932, the lowest since 1915. Even so, the International Harvester Company introduced the F-20, F-12, and the industrial version of the McCormick-Deering 10-20. Obviously, the company was counting on an eventual market rebound. Those predictions would prove correct, although it wouldn't happen until the second half of the decade.

By 1932, International was the number one tractor manufacturer with John Deere number two and J. I. Case third. That order would not change radically for decades, as John Deere and International Harvester would remain at the top of the list right into the 1980s.

The Farmall would start to lose some market share in the 1930s as other manufacturers introduced tractors that used the tricycle design to permit cultivation. Oliver, Massey-Harris, Minneapolis-Moline, Case, and Allis-Chalmers all had row crop tractors by 1930. John Deere was not far behind, and introduced its own row crop tractor.

To try and stay a step ahead of the competition, IHC increased the Regular's horsepower and weight to create the F-30 and F-20, and then used a mix of engineering and production savvy and tit-for-tat imitation to build the F-12.

The F-12, and later the F-14, were the most significant of the F Series Farmalls. For the first time, the standard tread and cultivating tractors used the same basic power units and similar chassis. The practice was continued to modern times, as shared designs conserved design and production costs.

The 1930s saw IHC's market dominance challenged and the introduction of several new models spoiled by the ravages of the Great Depression. But IHC was more than up to the challenge. Rather than pull back and wait for times to improve, the company pressed its advantage with more powerful engines and innovative equipment. Tractors continued to sell well, and IHC and the Farmall continued to be on the cutting edge of tractor innovation and sales.

Farmall F-30

During the development of the original Farmall, which became known as the Regular when the F-30 appeared, an assortment of weights and sizes had been tried; the final version was essentially a medium-heavy version. The experimentation with a heavier model played a role in the development of the Farmall F-30, which was simply a larger version of the Farmall Regular.

The F-30 was also confirmation of IHC's worst fears and hopes; the Farmall dramatically slowed sales of the McCormick-Deering 15-30. The sales of the Farmall were so strong that IHC management was convinced to back the concept wholeheartedly, and continue with its successful new design.

The F-30 packed the punch of the 15-30 into a more maneuverable package. It also gave the farmers a reason to replace their McCormick-Deering 15-30s with a new IHC product, which the company had to consider a good thing. The tractor was priced at about $1,150, about $100 less than the 15-30.

The tractor featured a four-speed transmission, one more speed than the Farmall had. At 12 feet, 3 inches, the F-30 was nearly 2 feet longer. It turned tightly, like the original Farmall, but took a 3-foot-larger circumference to do so. Still, a turning radius of just over 17 feet was impressive

The Farmall developed into a complete line, from the big F-30 down to the little F-12. All three of these models were introduced into the struggling economy of the 1930s, a time when tractor sales for all manufacturers dropped from about 200,000 in 1930 to just 19,000 in 1932. It also marked a decade in which the industry was beginning to catch up to the Farmall and introduce competitive row crop tractors.

for a tractor of that size. The McCormick-Deering 15-30 turning radius was nearly double that, at 30 feet.

The F-30 weighed about 5,300 pounds, which was 300 pounds more than the 15-30 and nearly 2,000 pounds more than the Regular. The F-30 engine used a 4 1/4 x 5-inch bore and stroke, making it a shorter stroke and a smaller bore than early McCormick-Deering 15-30s. The F-30 engine ran at 1,150 rpm. The additional 150 rpm was enough to give the F-30 engine more horsepower than the 15-30, and the extra weight combined with the slightly more powerful engine resulted in a bit more pulling power in the field.

Early Power Increases

Mind you, the F-30 was originally proposed to be built with a slightly less powerful engine. One of the first decisions on the F-30 concerned an increased-power engine and increased coolant capacity, and that decision received final approval on May 1, 1931. The decision bears a handwritten note stipulating that the serial numbers for the more powerful engines would be AA501 and up. Also, construction of the new engines was to begin at Tractor Works on July 15, 1931. Considering that only 623 F-30s were built in 1931, it's doubtful that any were made with the less powerful engine.

The tractor was equipped with a power take-off and an oiled air filter. The tractor could burn distillate, kerosene or gasoline. The manifold could route exhaust gas to warm the Schebler carburetor so that it would burn kerosene or distillate. Actually, distillate was considered the most economical farm fuel by 1925. It didn't produce nearly as much power as gasoline, but it was much less expensive. Kerosene wasn't as economical as distillate, but some farmers still used it.

The F-30 engine had a water pump and thermostat. This replaced the shutters used on older models to regulate cooling. An IHC E4A magneto with automatic impulse coupling was used.

Farmall F-30 Production	
Year	Production
1931	683
1932	3,122
1933	1,222
1934	1,506
1935	3,375
1936	8,057
1937	8,502
1938	1,821
1939	1,020
1940	NA*
Total Prod.	29,526

*Some documentation has shown that F-30s were built into 1940. Production figures not available. Source: [add; double-check numbers!]

Several specialized versions of the F-30 were built, although information on them is somewhat sketchy. Perhaps the most desirable today is the cane tractor, which was a narrow, high-clearance model with a wide front end. Other versions included a narrow tread and wide-front-end narrow tread tractors.

Pneumatic Tires

In 1933, the company authorized pneumatic tires as an option on IHC tractors. The rear tires for the F-30 were 11.25 x 24s (attachment number 25714-D) and the front tires were 6.00 x 16s (attachment number 25715-D). Pneumatic tires were a solution to a problem that was becoming increasingly troublesome for the farmer. Paved roads were becoming more common in rural areas, and lugged steel wheels severely damaged pavement. Laws were passed prohibiting the use of lugged steel wheels on paved roads, and farmers were forced to use planks anytime they needed to cross a paved surface.

Rubber-coated steel bands known as overtires and easily detachable lugs were attempts to solve this problem, but installing these devices was too time-consuming to be practical. Solid rubber tires were in use, mainly on industrial tractors, but they provided poor traction in wet conditions. Chains were a partial solution to this problem, but then the farmer was back to dealing with something that needed to be removed to cross pavement.

B.F. Goodrich experimented with rubber tires that were arched rather than air-filled in 1931, but Firestone provided the solution in the form of the pneumatic—air-inflated rubber—tire. The Firestone tires were introduced on Allis-Chalmer's tractors in 1932, and International demonstrated the same tires on IHC machines that same year.

The pneumatic tires provided good traction in all sorts of conditions, and had the added advantage of an improved ride for the tractor. In a 1934 test of lugged steel and pneumatic tires, the University of Nebraska Tests found that steel had a slight edge in mowing alfalfa and grass, but pneumatic tires worked more acres per hour while plowing, cultivating, combining, and harvesting crops. Conversely, steel had a slight advantage in fuel economy.

Another advantage of the pneumatic tires was the ability to fill them with water to weight the rear end for traction. Rear wheel weights were commonly used, and water was a simple way to increase rear traction without lugging heavy weights about.

In 1945, the Lift-All was adapted to the F-30 and offered as a retrofit kit with Cylinder Type Hydraulic Lift number 494015891.

The F-30 proved to be a rugged, maneuverable tractor, and did well for both IHC and the farmer. It was built and sold in respectable numbers until 1939 (with perhaps a few trickling out of the factory in 1940). The tractor slipped out of production when the Farmall tractor line was completely redesigned and the styled Letter Series was introduced.

When compared to its standard tread line-mate, the W-40, the F-30 outsold it at a two-to-one ratio every year except 1938. The F-30 proved that the Farmall design had merits in a larger, more powerful tractor. Although heavy field work had been the territory of the standard tread tractor—and still was with a certain segment of people—the big F-30 was proof that a tricycle tractor could work a field hard and retain the design's ability to straddle rows for cultivating and maneuverability advantages.

Farmall F-20

The next tractor to take the Farmall mold was the F-20, which was an upgrade of the Regular. The F-20 had more power than the Regular, and a few more features, and was introduced in 1932, the same year in which Regular production was reduced. There were apparently some complaints about the initial release of the F-20, and it was upgraded with more power and some additional features in 1936.

The F-20's transmission used four speeds rather than the three of the Regular, and the engine used a bore of 3 3/4 inches and a stroke of 5 inches. The four-cylinder engine was governed to run at 1,200 rpm and used removable sleeves, ball bearings on the crankshaft, and a combination of splash and gear-driven pump lubrication. It also had an oiled air filter. The engine was started on gasoline and could burn either kerosene or distillate. It was cooled with a thermosyphon system with a radiator. The clutch was an 11-inch single disk.

Farmall F-20 Production

Year	Production
1932	2,500
1933	3,380*
1934	662*
1935	26,334*
1936	36,033*
1937	35,676
1938	25,268*
1939	13,111*
Total Prod.	148,690

*Number of serial numbers issued that year, which is only an approximation of production numbers. Source: McCormick/IHC Archives, 1932 from "Tractor Production Schedule," 1937 and total from 150 Years of International Harvester

The F-20 used an engine similar to that of the Regular with the addition of new intake and exhaust manifolds and a new Zenith carburetor. The F-20 weighed about 4,500 pounds, just a few hundred pounds more than the Regular.

In August of 1938, the F-20 Cane Tractor was authorized to be built. The tractor was a narrow-tread F-20 with the addition of a high-arch wide-tread front axle with new steering knuckles, steering knuckle arms, and drag link in the front as well as larger front and rear wheels and foot brakes. The F-20 Cane Tractor was manufactured for use with tall crops. The company decision lists Australia as a key market for this tractor. Interestingly, this tractor could be equipped with a high-clearance attachment, which could be bolted to the front end of the F-20 Cane Tractor. This attachment increased front axle clearance by more than 12 inches, and widened the front axle tread nearly 6 inches.

Farmall F-12

The F-12 is one of the most significant members of the Farmall line, for two reasons. First, it came without the heavy and expensive-to-produce rear drive mechanism from previous Farmalls. Second, the F-12 was the first to be used as a basis for both the wheel and cultivating tractors. From the F-12 forward, International wheel and cultivating tractors would become increasingly similar. By using one basic model,

Sectional View of McCormick-Deering F-12 Tractor

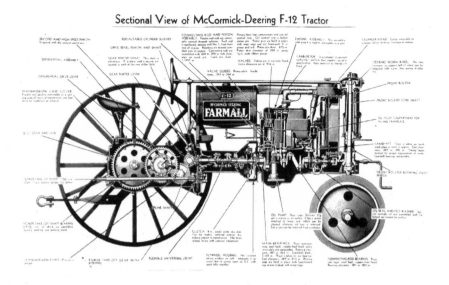

The F-12 was the fourth Farmall tractor, and was first produced in quantity in 1933, although experimentals and pre-production models were released early in 1932 and regular production models were released in October 1932. The earliest F-12s used Waukesha engines, although Harvester quickly switched to an IHC unit.

production costs were cut, engineering time was reduced, and retooling costs were slashed. By the 1950s, the limited difference between the standard tread and cultivating tractor became more subtle than in the past. Converting a row crop to a standard tread tractor required only a few parts to adapt the wide front end and lower the tractor a few inches.

The Regular used a dropped-rear drive mechanism. On each wheel, an enclosed, oil-bathed drive chain ran from the rear final drive down to the rear hub. This added complexity to the tractor, but increased ground clearance for cultivation. One of the drawbacks was that tread width was adjustable only by flipping the offset rear wheels.

The engineering stroke of genius that eliminated the dropped rear drive was simple, not to mention the fact that it was someone else's idea. The Case Model CC and Oliver Hart-Parr Row Crop used larger rear wheels to maintain the ground clearance necessary for cultivating. These tractors also made more power—about 18 horsepower at the drawbar and 28 on the belt—than all of the Farmalls except the F-30. For IHC to compete, it needed to eliminate the dropped-rear drive, and it did so with the simple solution of larger rear wheels.

This not only eliminated the dropped-rear drive mechanism, it also freed them from the restraints of a fixed tread width. Rather than fooling with clumsy reversible hubs, the rear axle of the F-12 was splined so that the wheel could slide freely. The tread width adjusted between 44 and 78 inches. This eliminated the need to produce a narrow tread version of the tractor, which was an expensive way to meet the farmer's needs.

The new concepts aside, the F-12's roots resided in early Farmall development. At one point, a lighter version of the Farmall was discussed. In May of 1921, engineer Bert R. Benjamin wrote company executive Alexander Legge, suggesting that they build a tractor similar in size and power to the Fordson with a narrow front end. Benjamin's contention was that such a machine could plow as well as cultivate. At the time, it would cost the farmer at least $1,200 to purchase both a tractor to plow and a tractor to cultivate. Benjamin felt the new IHC machine could be sold for about $700. He also felt that someone, probably Ford or International, would produce the tractor in the near future.

Benjamin's description was close to that of the F-12, a lighter tractor, weighing around 3,000 pounds, that sold for about $800 when introduced. These specifications closely mirrored those of the early Farmall designs in weight and power.

The company was enthusiastic about the new F-12, and felt it had more market potential than any of the previous Farmalls. The F-12 could be sold more cheaply than previous Farmalls, and Benjamin felt it was more profitable to operate than either animal power or the combination of a plowing tractor (typically a Fordson) and a cultivating tractor.

Early Production

In 1932, 25 pre-production F-12s were built and sold. Quite a few of them went to Texas, but the rest were sprinkled around the Midwest and Southeast, with at least one going west to Arizona.

The early models used a Waukesha Model FL four-cylinder engine. Later models used an IHC engine. The engine, with a bore of 3 inches and a stroke of 4 inches, turned at 1,400 rpm, and an oil filter and pump were used. The carburetor was a Zenith 93-1/2W, cooling was by thermosyphon, presumably with a radiator, and the clutch was an 8-inch single plate. The transmission was a three-speed, and a PTO and belt pulley were standard equipment. Also standard were an IHC oil air filter and spade lugs.

Farmall F-12 and F-14 Production

Year	Production
1932	25
1933	4,355
1934	12,530*
1935	31,249*
1936	33,177*
1937	35,681*
1938	6,425*
Total Prod.	123,442

Note: 1932 and early 1933 F-12s used the Waukesha Model FL four-cylinder engine. Subsequent machines used an IHC four-cylinder engine.

F-14

Year	Production
1938	15,607*
1939	16,296*
Total Prod.	31,903

*Number of serial numbers issued that year, which is only an approximation of production numbers. Source: McCormick/IHC Archives, 1932 and 1933 from "Tractor Production Schedule"

The tractors must have performed acceptably, because 1,500 more of the same machine were authorized to be built in early December of 1932. Either some kind of problem was discovered with the Waukesha engine or IHC wanted to increase profit margin, because the Waukesha engine was jettisoned only a few weeks later. Late in December of 1932, the decision was made to build a new IHC four-cylinder for the F-12. The engine had a bore of 3 inches and a stroke of 4 inches, ran at 1,400 rpm, and was cooled by a thermosyphon system. The engine was equipped with a new IHC 1-inch down-draft carburetor, a Purolator oil filter, and an IHC magneto. The original decision specifies the IHC E4A magneto with a handwritten note stating that the IHC FA magneto was actually used on the F-12.

Later references to the F-12 and F-14 indicate that both gasoline and kerosene versions of the tractor were made. It does make some sense that the dual-fuel manifold would be left off the machine to help keep the cost down. Use of the IHC engine required extensive changes to the control levers, cooling system hookup, and exhaust and intake attachments. The main frame beams were changed as well. A total of 30 changes were listed on the specifications sheet, undoubtedly requiring a fair amount of retooling and retrofitting.

New Engine for 1933

So, the F-12 sported a new engine for its 1933 release, and sold beyond the 1,500 allotted to be built that year. Over 4,300 F-12s were built in 1933, which was a modestly successful first year. Sales picked up from there, and the F-12 sold more than 35,000 units in 1937, which was a strong number for the time. Certainly IHC was turning a profit on the

The F-12 was an evolution of the Farmall tractor. Early production models were equipped with Waukesha engines.

little F-12, but the numbers were not perhaps as strong as management had hoped. It was successful, but hardly more so than the fast-selling Farmall Regular.

F-14

In 1938, IHC decided to see if giving the F-12 a little more power would put more life into sales. The engine was governed at 1,650 rather than 1,400 rpm, and the gearing was adjusted to maintain the same vehicle speeds. Some additional improvements were made, including raising the seat and steering wheel to a higher and purportedly more comfortable position, adding a heavier seat spring, and moving the clutch and hand brake levers to accommodate the repositioned seat and steering wheel. Also, a new PTO pinion gear was installed.

12 Series Production

Year	Production
O-12 and O-14	
1934	580*
1935	534*
1936	651*
1937	984*
1938	621*
1939	406*
Subtotal	3,776
Total O-12/ O-14 Prod.	4,793
I-12	
1934	263*
1935	200*
1936	550*
1937	970*
1938	209*
Subtotal	2,192
I-14	
1938	310*
1939	585*
Subtotal	895
Total I-12 and I-14 Prod.	3,087
W-12	
1934	853*
1935	675*
1936	737*
1937	1,031*
1938	334*
Subtotal	3,630
W-14	
1938	476*
1939	686*
Subtotal	1,162
Total W-12 and W-14 Prod.	4,793

Number of serial numbers issued that year, which is only an approximation of production numbers.
Source: McCormick/IHC Archives

In February of 1938, it was determined that the modified F-12 would be known as the F-14. A belt pulley was added to the list of standard equipment, and the new F-14 was introduced in mid-1938. The decision notes that all F-12 accessories and implements would work on the F-14, with the exception of the hydraulic lift attachment. The hydraulic lift attachment for the F-14 was different than the unit available for the F-12.

Electrics

An electric starting and lighting package for the F-12 and F-14 was introduced in 1937, and modified in 1938. The modifications allowed the battery box to be located on either side of the tractor to accommodate implements. Two different part numbers were used for the kits for gas or kerosene models, but the sole difference between the kits was the choke and governor control brackets.

In 1939, the F-14 sold more than 16,000 units, a respectable number. The tractor had carved out yet another niche for IHC tractors, proving that the market would bear a smaller general

Paint Transitions

The Day International Saw Red

A long-time debate centered around the exact date International Harvester Company (IHC) tractors were first painted red. An article in *Red Power* magazine stated that its findings indicated the date was November 1, 1936, which turns out to be right on the money. Cindy Knight, the archivist for the McCormick-International Harvester Collection at the University of Wisconsin, and researcher Guy Fay, located the official decision in 1995, and the mystery was solved (of course, those of you who read *Red Power* aren't surprised).

Beginning with tractors built on November 1, 1936, IHC tractors, crawlers, and power units were painted red rather than gray. This came at a time when tractors of most brands were switching from drab grays and olive greens to brighter colors, and IHC gladly jumped on the bandwagon.

The decision specifies that the entire wheel tractor except the ground wheels be painted "Harvester No. 50 motor truck synthetic enamel." The wheels were to be painted "Harvester red color varnish." Ditzler 71310 is generally accepted as the correct red paint for International tractors.

For crawlers, the tracks and track frames were painted with Harvester red color varnish, and the rest was painted with Harvester red No. 50. Power units were painted entirely Harvester Red. Modified industrial tractors were supposed to be painted Harvester red as well.

Note, however, that industrial tractors would be painted to the specifications of the purchaser. Although the wording of the decision refers only to industrial tractors, IHC researchers feel that all tractors could be ordered in special colors. Certainly white and yellow Letter Series tractors were built. Whether tractors were painted other colors at the factory is one of those mysteries that keeps a certain type of hard-core enthusiast awake at night.

The W-12 marked the beginning of a trend for International. While the W-30 bore little or no resemblance to the F-30, the W-12 and F-12 shared the same basic chassis. The result was reduced costs on everything from production to parts.

purpose tractor. The F-12 and F-14 were also the parental units of two of today's most popular, collectible, and useful IHC tractors, the Model A and Model B. Those tractors would appear in 1939, carrying a new look that continued IHC's dominance in tractors.

Standard Tread 12s

The standard tread 12s, as the W-12, I-12, O-12, and Fairway 12 are referred to in this book, were a significant step forward for IHC. From almost the first day IHC began producing tractors, several different divisions or factories built an assortment of similar tractors with different engineering bases. The result was independently engineered and produced machines, requiring different tooling, independent development costs, and non-interchangeable parts. This drove up costs and, in some cases, the tractor lines competed against each other.

Very early in IHC history, all of this was understandable because divisions were formed from independent companies. The animosity between the makers of the Titan and Mogul lines may have been a natural result of the people having battled tooth-and-nail one day and then becoming company bed fellows the next. The internal competition may have spurred some designs, and perhaps even created the Farmall, but the process was not cost-efficient.

With the standard tread and F-12s, IHC could focus on building one basic layout, and fairly easily modify it for both the cultivating and wheel tractors. One engine. One basic frame. Interchangeable parts. Costs down, price down, and, theoretically, sales going up.

The company's theory was correct, and its predictions bore fruit. The standard tread 12s did not sell as wildly as the F-12s, but IHC did not have as many developmental dollars to recover. Modest sales would return a profit, which was exactly why IHC wanted to move to a single base engine and chassis design. Note, however, that the W-12 was still built in a separate facility—Tractor Works—while the F-12 was built at Farmall Works. Despite all the company's efforts to bring the two machines to one common base, they were still built on two different production lines in two separate locations. Unified production sites would have to wait for the Letter Series.

Regular Production Begins

The field, orchard, and industrial versions of the little 12s were authorized on the same day, December 4, 1933. They were introduced as 1934 models, and all three used the same engine as the F-12, with a slightly different frame.

These tractors weighed about 3,200 pounds, just like the F-12, and used hand-operated brakes and throttle, a three-speed transmission, and an 8-inch single-disk clutch. The frame was a two-piece unit bolted together in the middle. The front axle was a low-slung unit with tapered roller bearings for the wheels, and the rear axle was semi-floating with diaphragm oil seals. Rubber tires were standard, and the cooling system was thermosyphon with shutters on the radiator controlling temperature. A swinging drawbar and pan seat completed the package.

The orchard version bore swoopy fenders and low-slung exhaust and intake pipes, while the industrial had different brakes and drawbar. Both wore rubber tires, while the I-12 had foot brakes and the O-12 brakes were hand-operated. The transmissions in the I-12 and O-12 had a tall third gear, which allowed the little tractor to sprint through the orchard or across the warehouse at just over 10 miles per hour. The W-12 rode on steel wheels, had a farm-type drawbar, and was geared a bit lower for field work.

The fourth addition to the standard tread 12s was the Fairway 12, which was simply an O-12 with steel wheels and an I-12 foot brake. The Fairway was designed for—you guessed it—use on golf courses. The Fairway 12 was authorized in April 1934.

In 1936, the 12s were upgraded a bit. The 8-inch clutch was replaced with a 9-inch unit, an agricultural-type seat was fitted, foot brakes replaced the hand brake, and the wheels and tires were changed.

The front half of the frame was upgraded in 1937 to accommodate electric starting and lighting. The flywheel, steering gear housing, rear engine support, timing pin, and clutch pedal return spring were new, in addition to the new front main frame.

The little 12s provided International with an easily convertible wheel tractor, and made it easier for the company to offer such machines.

Chapter Six

The Industry of Design

Farmall Models A, B, and C

"I once said that the most difficult things to design are the simplest. For instance, to improve the form of a scalpel or a needle is extremely difficult, if not impossible. To improve the appearance of a threshing machine is easy. There are so many components on which one can work."
—Raymond Loewy

As the 1930s drew to a close, the competition was beginning to nip at the heels of the International Harvester Company. The company was the number one producer, but John Deere and Case were both gathering steam. These companies, along with a host of others such as Oliver, Massey-Harris, and Minneapolis-Moline, had row-crop models to compete with the Farmalls. In addition, the farm market slumped a bit in 1938, and IHC's earnings dropped about 33 percent. The climate was not as hostile as the tractor wars of the 1920s, mainly because the use of the horse was finally declining and tractor sales were booming. All the same, IHC needed another stroke of genius to stay on top.

The Letter Series tractors provided just that. The tractors sported innovative features combined with a functional, streamlined look. The gangly appearance of the earlier Farmalls was replaced with the graceful design of Raymond Loewy, one of the era's most prominent industrial designers. The new line was part of a company-wide design move that included a new logo. The new machines were produced for more than a decade, and the basic form endured until the 1960s, when the round lines of International tractors were replaced with the squared-off look that lasted to modern times. The logo survived until 1985, when the IHC tractor division was merged with the Case tractor division.

The Letter Series innovation stemmed more from refinement than dramatic engineering breakthroughs, but the end result was revolutionary. The key to the tractor's new design was Raymond Loewy. At the time, industrial designers were changing the face of all kinds of products. Everything from toasters to pop machines to diners were being streamlined and styled. The result was big sales, but the concept represented a fundamental change in the way machines were designed.

In the early part of the twentieth century, new technology dramatically changed the way America worked and lived. The automobile and tractor were two of the most obvious innovations, but every facet of people's lives was affected by the rise of power equipment. Manufacturers concentrated on efficient, effective equipment. The tractor changed from a difficult, clumsy beast to an efficient tool. Similar evolutions were reflected in all types of equipment.

The evolution of the machine brought more efficiency and reliability, but the machines were designed for tasks rather than for people. The approach could be seen from all angles. Machines were harsh and jarring in design and typically unfriendly and uncomfortable to use.

What Loewy and his contemporaries brought to the table was more than streamlined toasters and coffee makers. It was the idea that machines should be designed for people to use. The success of the industrial design movement can probably be credited to the fact that the new machines looked futuristic and exciting. But the significance of the movement was that the machines were shaped to humans rather than vice versa.

Bert R. Benjamin brought in a new revolution by concentrating his engineering efforts on the task rather than the machinery. Loewy helped IHC bring another revolution by concentrating on the operator. The result is a series of timelessly attractive tractors that retain their utility 60 years after their introduction.

The smaller Farmalls were designed for the small farm, and were the most likely to directly replace a horse. By the 1940s, the tractor had turned the corner and more farms used tractors than not. Still, there was a reasonably large contingent of small, low-profit operations using horses. The initial cost of a tractor was one barrier for these farmers, who didn't necessarily have the capital or good standing with the bank to purchase a new machine. Profitability was another factor. A tractor that was too large to perform most of the farm work wouldn't save money even if the farmer had enough money to buy a tractor.

The Model A, B, and C addressed those very concerns. The tractors were economical to run and purchase, agile around the farm, and able to perform a wide variety of tasks. The Model A and B appeared in 1939, debuting with "Cultivision" and a complete line of implements. The Model C appeared in 1948 as a replacement for the Model B. The A and C both went "Super" near the end of their production runs, with boosted power and new features. As far as IHC tractors were concerned, these machines were super throughout their lifetime, with impressive sales figures and hundreds of thousands of satisfied Farmall owners to their credit.

Farmall Model A and Model B

The Model A and Model B were the lighter and more powerful replacements for the Farmall F-14. In many ways, they were the most innovative of the first group of Letter Series tractors. As with all of the new tractors that year, the models used streamlined sheet metal and smooth lines. The gangly appearance of the previous Farmalls was transformed into a much sleeker, more attractive package.

Beyond the new sheet metal, the most distinctive feature was the off-set of the operator and the engine, which truly reflected the industrial design concept. Think about the operator first. Ground clearance and tread width are key to cultivating, but neither of these things is worthwhile unless the operator can see what he or she is doing. The result was what IHC called "Cultivision." The engine was offset to the operator's left, and the view to the ground was unobstructed.

The tractors also represented a union between IHC's desire to use Henry Ford's manufacturing techniques and the revolutionary Farmall design pioneered by Bert R. Benjamin and company. The roots of their design harked back to Benjamin's suggestion to build a light tractor like the Fordson that was capable of cultivating as well as plowing. The Model A and B were just that, with enough power and weight for light plowing, and adjustable tread width and great vision for cultivating.

More significantly, the two models used the integral frame and unit construction developed for the McCormick-Deering 15-30 and 10-20. The engine, steering gear housing, clutch housing, transmission case, differential housing, and final drive housing were modular pieces. The pieces were then bolted together to effectively become the tractor's frame. The design was strong, efficient and reasonably easy to service.

Engine and Transmission

The four-cylinder engine had a bore of 3 inches and a stroke of 4 inches. The engine was rated at 1,400 rpm, but was controlled with a variable speed governor. The engine was lubricated by an oil pump using an oil filter with a removable star-shaped element. Ignition was an IHC H4 magneto with automatic impulse coupling and an integral grounding switch. The carburetor was an updraft Zenith Model 61AX7 with a 7/8-inch throat, and the tractor was designed to burn gasoline. The compression ratio was 5.33:1, which was considered sufficient to burn gasoline. The engine was cooled with a closed thermosyphon system and a fan. Later versions of the Super A used a regular water pump and thermostat.

The clutch was a 9-inch single plate operated by a foot pedal. The transmission had four forward speeds, and used 11 roller bearings and 1 ball bearing. One advantage of the increased power was that it allowed a broader range of transmission gears. In top gear, the Model A would pull a blistering 9 3/4 miles per hour. This was a significant increase from the top speeds of previous tractors, which were limited to 3 or 4 miles per hour. Kits had been offered for several years to install a "road gear" to bring tractor speeds up a bit. The Model A was one of the first to finally receive a top gear suitable for moving the tractor from place to place.

Chassis and Final Drive

Physically, the tractors were small and light. The operating weight of the Model A was just under 1,700 pounds, more than 1,000 pounds lighter than the preceding F-14. The Model A was shorter and a bit narrower than the F-14, as well.

One of the oddest engineering decisions on the Model A and B was the dropped-rear drive mechanism. The simple straight axle and larger rear wheels introduced to IHC on the F-12 provided the advantages of easily adjustable rear wheel tread and reduced complexity. The Model A and B used the same dropped-rear drive found on the Farmall, and experienced the same problems. Rear tread width could only be adjusted by flipping the offset rear wheels and the additional parts added weight and complexity. It's possible that IHC wanted to keep rubber costs down, as using the mechanisms allowed smaller rear tires to be mounted. Perhaps the added weight of the mechanisms were intended to improve traction.

Equipment, Standard and Optional

The standard wide front end on the Model A was not adjustable. An optional adjustable front end provided eight settings that ranged in 4-inch increments from 40 inches to 68 inches.

Pneumatic tires were widely used by 1939, and were standard equipment on the Model A and B. All of the Model As used a wide front end and were designed to cultivate a single row. For those who wanted a two-row cultivator, the narrow front-end Model B and BN were offered. The turning radius on the wide front models was 9 feet.

The brakes were external contracting bands on forged steel drums mounted on differential shafts in housings. They were operated by foot pedals that are depressed separately for turning or together for braking.

The Model A was part of the Letter Series, the most successful International tractor line of all time. Over one million of these streamlined Farmalls were sold between 1939 and 1954. The crawler is IHC's largest built at the time, the TD-18.

The rubber on the wheels smoothed the ride out a bit, and the new seat helped a bit more. The pan-style steel seat of past machines was discarded for a padded seat covered with cotton duck mounted on a spring. The seat was adjustable, and could be tipped back to accommodate standing. The seat was another reflection of the design of Raymond Loewy. Function was king, and a machine that is operated for hours on end should be comfortable. The advance was undoubtedly well-received.

Power Lifts

Power lifts were becoming increasingly popular on tractors by the 1940s, and the Model A and B had several options. One was a simple hand-operated mechanical lift. If the A or B owner wanted a power lift, the optional pneumatic lift was the answer. In many ways, the lift was an elegant engineering solution, although it had one major drawback. The lift was powered by exhaust pressure, and provided plenty of lift power without the drain on engine power of a hydraulic lift. The drawback was that if the mechanism got dirty (which tends to happen in farm conditions or after 20 or more years of sitting idle), the lift became less than powerful. Regular maintenance kept the pneumatic lift working well, but neglect would bring only misery. The Super A had a hydraulic system to power the lift, and the pneumatic lift was phased out. Although the hydraulics required less maintenance, the hydraulic system soaked up horsepower. Some say that the Model A actually has more drawbar power than the Super A due to this.

Other attachments available for the Model A included a swinging drawbar, belt pulley, power take-off, wheel weights, exhaust muffler, equipment for burning distillate and kerosene, electric lighting, a radiator shutter, and several different pistons to change the compression ratio for use at higher elevations. Shortly after being introduced, the Model A received a redesigned rear wheel and a PTO shield.

Farmall Model B

The Model B was almost identical to the Model A, but was designed to cultivate two rows at a time. The single front wheel went between the rows, while the dual front wheels of the Model A straddled a single row. The B also had a wider rear tread width and a tricycle front end. The Model A and B used the same series of serial numbers with different codes.

This Farmall B is a custom-built machine that features a wide front rather than the standard narrow front end. The factory never made these.

The rear tread width on the Model B was adjustable from 64 to 92 inches, compared to the range of 40 to 68 inches of the Model A. The wider rear end made the B slightly heavier (just under 1,800 pounds). The same attachments were offered for the Model B as for the Model A.

New Engine Package for Model A and B

In 1940, a high-compression package was authorized for the Model A and B. The company decision (a dated company document ordering engineering actions) reads that the option was offered to meet sales demand and to "provide equipment as tested at Nebraska." The wording insinuates that the University of Nebraska may have tested a more powerful Model A or B than the version IHC released to its dealers. Detroit auto makers have certainly been willing to release factory-massaged machines for testing, so it doesn't seem a huge stretch that IHC would send a more powerful machine to the Nebraska test. The fact that the decision was not to be circulated to distributors or even the head office makes it fairly likely that IHC pulled a fast one on the Nebraska tractor test of 1939.

Whatever the cause, the high-compression package was offered late in 1940 or early in 1941. It consisted of a cylinder head, valve guide, valves, seats, intake and exhaust pipe, and all the necessary gaskets, studs, and brackets. No horsepower figures were on the decision, but it's safe to assume that tractors equipped with the package would match the output of the Model A tested in 1939. The output of tractors not equipped with the package is unknown.

Farmall Model BN

In August of 1940, a new version of the Model B was authorized, the Model BN. The tractor was simply a Model B with a narrower rear tread width. The tread width was adjustable in 4-inch increments from 56 to 84 inches by ordering optional rims with different amounts of offset. According to the factory decision, the parts required to convert a standard Model B to a Model BN were a new left differential shaft, differential housing, platform, and drawbar. The right seat support on the Model BN touches the right fender, while the Model B has a gap.

Collector Jim Becker notes that the Model BN has several advantages over the Model B. For one, fewer were built, making it a rarer model.

The Farmall B was the two-row counterpart to the one-row Model A. The major difference between the two was the narrow front end, which was designed to ride in the middle of two rows. This Texas farmer is using a B-236 cultivator.

More importantly, the slightly narrower Model BN fits into his trailer, while a Model B will not.

A swinging drawbar was available as an attachment. A single front wheel was standard, and the more common closely-spaced dual front wheels were optional.

The BN was narrow and light for a tricycle design tractor, and was a bit tippy. It provided an expensive solution for the farmer who worked tightly spaced crops, but tread width adjustment was still a pain. These problems undoubtedly influenced the development of the Model C, which was heavier, wider, and had the easily adjustable straight axles of the larger Letter Series tractors.

Farmall Model AV

The Model AV was the high-clearance version of the Model A. It was intended for use in tending to crops planted on high beds that required additional tractor ground clearance. Sugar cane was the most common crop requiring a high-clearance machine, and cane fields used the higher beds to improve drainage.

The Model AV consisted of a group of additions to the Model A. It was authorized for production in October of 1940. The package consisted of a new front axle, front wheels, rear axle shafts, rear wheels, drawbar, and steering gear housing base. The result was a 5-inch increase in ground clearance, which brought it up to 27 1/2 inches. The front wheels were 4.00 x 19s, and the rears were 8.00 x 36s. The transmission gearing was not changed to accommodate the larger wheels, so the tractor traveled slightly faster in each gear than a Model A. Top speed bumped up about 3 miles per hour to 12 3/4 miles per hour.

Available attachments included a swinging drawbar, wheel weights, belt pulley, PTO, muffler, equipment for burning distillate and kerosene, radiator shutters, electric starting, electric lights, flywheel ring gear, double-groove fan and generator pulley hub, spark arrestor, tire pumps, heavy duty front axle, high-compression gasoline starting, and different compression ratio pistons for higher altitudes.

International A

In 1940, the International Model A was introduced. The International was the industrial version of the Model A, and it closely resembled the Farmall Model A. The industrial models were intended for commercial use, and researcher Jim Becker feels that most of them were used as highway mowers. The tractors were certainly marketed to different audiences, and the International Model A shows up in all kinds of situations—from orchards to factory applications—in archival photographs.

According to Becker, the major differences between the International A and Farmall A were the International A's foot accelerator and the Model A's optional heavy-duty front axle. The diamond-pattern pneumatic rear tires were more common on Internationals, although they became available for any of the Model As.

The convention of stickering industrial or utility versions as Internationals and the cultivating version as Farmalls eventually became common IHC practice. It was unique to the Model A during Letter Series production, but the Hundred Series would use the same convention, replacing the W, I, and O tractors with a standard tread model badged as an International. Orchard and industrial versions could still be ordered, but they no longer had a unique model designation.

Farmall Super A

The Super A was an upgraded version of the Model A that featured a more powerful engine mated with a host of optional equipment that included Touch Control hydraulics, quick-attach implements, and the universal frame mounting system.

Touch Control

Touch Control hydraulics debuted on the Super A and the Model C. Touch Control was a hydraulic lift that powered implements up as well as down, giving the operator a powered lift at the touch of a lever. Lifts had been around for quite some time, but hydraulics were an up-and-coming development. They eliminated the need for clumsy mechanical linkages to a lift, and afforded great mobility to tractor attachments. Unfortunately, they also required horsepower to run. The horsepower devoted to hydraulics was lost, and the tractor was a bit less efficient at the drawbar. Some even say that a Model A had more pulling power than the Super A due to the horsepower lost to the Super's hydraulic system. Later Touch Control systems had a temperature gauge mounted between the control levers to monitor hydraulic fluid condition. This gauge was used on the Super C and perhaps the Model C as well.

Decals

Super A decals are another interesting topic. Several versions of the decal sets are floating around, with quite a bit of confusion on what was actually used and when. The first issue is the "McCormick-Deering" lettering on the hood. Early Letter Series tractors use "McCormick-Deering," while later machines tend to have just "McCormick." Researcher Jim Becker has found that original International documentation shows that Deering began dropping off the logo in 1949, although the exact date varied from model to model. Note that the address listed below "Farmall," also changed, probably about the same time as "Deering" disappeared.

Two distinctly different "Super A" logos were used. The type commonly found on archival photographs is a white circle around a white letter "A" with a wavy script "Super" banner behind the letter "A." The other logo is a filled white circle containing a red letter "A" and red "Super" text in a semicircle within the circular logo and above the letter "A." The second is more commonly seen today, and is the decal most often found for sale. The first, with the wavy "Super" type, is the one Becker feels the factory used on all Super As. Archival photography seems to support his belief, but many other IHC researchers believe the typical round white logo was used on factory machines as well as the wavy type.

Fast Hitch

International's two-point hitch system, dubbed Fast Hitch, was also available on the Super A. This hitch was IHC's answer to the Ferguson three-point hitch. Fast Hitch was a high-quality system, and its advantage over the Ferguson system was the ease of mounting. The operator could simply back the tractor into the implement and go, rather than compete in the wrestling match involved in attaching an implement to Ferguson's three-point system. The drawbacks (for the end user) were that Fast Hitch implements had to be purchased to take advantage of the new system, and the implements could not be used on other brands of tractors.

Universal Mounting System

The universal frame mounting system was another advance designed to simplify the farmer's life. Attaching implements to a tractor's frame typically involved bolting on an attachment frame. Each implement required a unique frame, making changing implements a major pain.

With the universal frame mounting system, all IHC attachments used the same mounting frame, and switching implements became easier. Fast Hitch actually provided a similar service, as all Fast Hitch implements used the same mounting system on the tractor.

International A (Super)

The International version of the Super A was still known as the International A. It featured all the new Super A bits, plus a beefier Touch Control system with heavier control arms and slightly more powerful internals.

Model A and B Production

Model A, AV, International A, Super A, Super AV, and Super A-1	
Year	Production
1939	NA*
1940	22,023
1941	22,950 Ї
1942	9,579
1943	105
1944	8,177
1945	18,494
1946	19,739
1947	20,937 §
1948	15,869
1949	13,805
1950	16,376
1951	27,562
1952	11,334
1953	17,909
1954	5,953†
Total Prod.	230,812

*Although the tractor was definitely produced in 1939, the Tractor Production Schedule lists first production in 1940. This is probably just an accounting issue, as the 1939 machines may have been considered 1940 models by the company.
Ї International A introduced
§ Super A introduced
†Super A-1 introduced; about 2,000 Super A-1 serial numbers issued in 1954
Source: McCormick/IHC Archives, "Tractor Production Schedule"

Farmall Super A-1

The Super A-1 used a larger bore to bring displacement of the four-cylinder engine up to 123 cubic inches. Also, it had a thermostat to go with the water pump, bringing the cooling system up to date. The Super A-1 was produced only in 1954. About 2,000 serial numbers were issued that year, so production should have been close to that.

The White Letter Series Tractors

The white Letter Series tractors were built as part of a promotional campaign in 1950. An article by Darrel Darst in *Harvester Highlights*, the newsletter of the IHC club, states that white demonstrators were built for three months at the Louisville Plant in Kentucky. The entire line of equipment built at Louisville was painted white.

Farmall Model C

Introduced in 1948, the Model C was the replacement for the Model B. It was larger and heavier than the Model B, but used the same narrow front end and was designed to be a two-row cultivator. Like the Model A and B, the Model C was designed for the small farmer who was perhaps upgrading from horses to power farming.

The Model C no longer used the dropped-rear drive of the Model A and B. Instead, the Model C used larger rear wheels and a straight axle like those introduced on the F-12 and used on the Model H and M.

The straight rear axle allowed the rear tread width to be infinitely adjustable within a range, and tread width could be adjusted without taking the rear wheels off. It required larger rear tires, but eliminated the extra tooling and complexity of the dropped-rear drive mechanism.

The Farmall C engine was the same 3 x 4-inch four-cylinder used in the Farmall A and B. By governing the engine to run 250 rpm higher, at 1,650 rpm, a couple of extra horsepower were coaxed from the Model C engine. Burning gasoline, output was about 19 horsepower at the belt pulley and 15 horsepower at the drawbar, enough to pull a two-bottom plow in average conditions with ease. The engine used an IHC magneto and a Zenith carburetor.

The engine was the same as that of the Model A and B, but the chassis was significantly larger and heavier. The Model C weighed about 3,000 pounds, significantly heavier than the 1,700 pounds or so of the Model A.

The chassis used the engine and four-speed transmission cases as stressed members, like the Model A and B. The front end bolted to the engine cases, and the rear used straight splined axles that allowed the tread width to be adjusted.

Farmall Model B and BN Production

Model B, BN	
1939	NA*
1940	12,765**
1941	16,553
1942	6,305
1943	5
1944	7,933
1945	12,951
1946	14,623
1947	20,100
1948	1,921
Total Prod.	93,156

Although 1939 serial numbers were issued, the Tractor Production Schedule lists production beginning in 1940
**Model BN introduced in 1940*
Source: McCormick/IHC Archives, "Tractor Production Schedule"

The Farmall C was introduced in 1948 as a replacement for the Model B. The Farmall C used a straight rear axle rather than the dropped-rear drive mechanism found on the Model A and B.

The Farmall C also had Touch Control hydraulics, like those on the Super A. Options included a wide tread model, although these are fairly rare. The Model C was quite popular, and outsold the Model A and Super A with production of more than 20,000 units per year.

Farmall Super C

The Super C was an increased power version of the Model C. The increased power came from a bit more displacement found with a larger bore, up from 3 x 4 to 3 1/8 x 4 inches. The engines were now rated by displacement rather than bore and stroke, and the four-cylinder gasoline-burning engine displaced 122.7 cubic inches. The horsepower increase was modest, with a gain of a couple horsepower at both the belt pulley and drawbar.

The Super C introduced International's new double-disk brakes. The disk brakes replaced the drum units, and were covered and sealed from grit. Fast Hitch later appeared on the Super C, along with a plug-in adapter that allowed the use of three-point hitch implements.

The Model A and B introduced the tractor to the farm for hundreds of thousands of people. The little tractors were small enough for the simplest farm task, and powerful enough to perform as real tractors. Actually, both made more horsepower than the original Farmall. The Model C was slightly larger, and performed

Model C and Super C Production

Year	Production
1948	15,547
1949	26,338
1950	24,280
1951	37,651†
1952	31,130
1953	29,472
1954	13,753
Total Prod.	178,171

†Super C introduced
Source: McCormick/IHC Archives, "Tractor Production Schedule"

the same function as the two-row Model B with the added advantage of easily adjustable tread width and a more stable chassis.

All three of these machines served well on the farms of the 1940s and 1950s, and many are still at work today. The tractors are ideal for a hobby farm, and just as useful on today's small farm as they were when they first appeared.

Chapter Seven

The Tractor Age

Farmall Model H and M

A pessimist is one who makes difficulties of his opportunities and an optimist is one who makes opportunities of his difficulties.
—Harry S. Truman

The 1940s and early 1950s were the golden age of the tractor. The horse was nearly gone, and the tractor was king. Farmers were working more land than ever before, and molding that land to suit their needs. Machine power was vital to the mid-century vision of successful farming.

For the farmer of 100 acres or more, the Model H and Model M were the ideal tool. The tractors were powerful jacks-of-all-trades, capable of pulling feed wagons and turning conveyor belts as well performing the heavy field work of plowing, harvesting, planting, and cultivating.

The early 1940s were incredibly successful for tractor manufacturers, especially International. Even World War II did not take the shine off the company during this era. Sales dipped in the mid-1940s, but nothing like the slump experienced during World War I. Savvy tractor manufacturers built crawlers for the war effort, and were able to convince the

The industrial design of Raymond Loewy added much more than style to International's new-for-1939 Letter Series tractors. His concept of building the tractor for the operator was applied from top to bottom, with lots of little touches designed to make the tractor easier to use.

government that tractor production was important enough not to divert material from their production. With the nation's men at war, the remaining farmers needed efficient power more than ever, so the tractor rolled on.

Early in the 1950s, with the post-war boom still in stride, IHC's sales of mid-sized and larger tractors would peak, never to reach that pinnacle again. The tractors sold in unprecedented numbers for a full-size tractor. The Model H alone sold nearly half a million units. The numbers were a result of the fact that farmers were turning to tractors like never before, and the Letter Series was one of the favorites of the time.

During the tail end of Letter Series production, the diesel engine would appear in cultivating tractors. The Super series machines would see the introduction of the independent power take-off, Torque Amplifier, and improved hydraulics. With these improvements, the fundamental functions of the modern farm tractor were in place.

Farmall Model H

The Model H replaced the F-20 with a stylish package that featured all of IHC's latest bells and whistles. The Model H and the Model M were based on the same chassis design. Although the two had the same mounting points for implements, they shared very few parts. The Model H was designed to be a two-bottom plow tractor while the Model M was rated to pull a three-bottom plow.

The Model H used modular construction, like the smaller Model A and B. Components could be taken off independently as an assembled unit. This had some advantages when the tractor was being built in the factory, and made servicing the tractor much simpler.

Chassis

The Model H's frame was a hybrid of an integral and beam frame. The rear of the frame was the transmission and final drive housing, while the front was a beam frame that ran from the clutch housing forward to underneath the engine.

The frame design showed touches of Loewy. The engine was rubber-mounted to the front frame channels, presumably to make the tractors smoother to operate. A little less vibration would make a big difference over a long day in the fields.

Engine

The Model H's four-cylinder I-head engine used a 3 3/8-inch bore and a 4 1/4-inch stroke and was rated at 1,650 rpm. The engine for the original version was started on gasoline and ran on distillate or kerosene (a gasoline-burning engine appeared in 1940). A variable-speed governor was used. The engine oil was force-fed by an oil pump, and an oil filter strained grit from the lubricant. Air and fuel cleaners were standard, as well. An IHC H-4 magneto and IHC D-10 carburetor were used. The distillate-burning version of the Model H was tested at Nebraska in 1939, and produced 22.14 horsepower at the belt pulley.

The engine was cooled with an enclosed system that used a radiator and water pump. Cooling was adjusted with shutters that could be controlled by a hand crank. Later models used a thermostatically controlled system, with shutters as an available option for cold-weather use.

Transmission and Clutch

The transmission used five speeds, with top gear yielding nearly 17 miles per hour. The top gear was used only with pneumatic tires and was blocked out on steel-wheeled machines. A few models were sold with steel when the tractor was introduced, and a significant number

The Silver Spoon Executive

Harold Fowler McCormick Jr.

Harold owler McCormick Jr. was born in 1898 as the heir of two of the country's most powerful dynasties, the McCormicks of the International Harvester Company and the Rockefellers of Standard Oil.

Best-known as simply Fowler, McCormick was the defintion of of a fortunate son. His grandfather, Cyrus McCormick, was the inventor of the McCormick Reaper. Fowler's father, Harold Fowler McCormick Sr., became the chairman of the board of IHC in 1935. Fowler's mother, Edith Rockefeller McCormick, was the daughter of John D. Rockefeller Sr., and an heiress to the Standard Oil fortune.

Fowler's father, Harold, was at best a bit of an eccentric. Harold divorced Edith Rockefeller and, while single, had monkey testicle parts transplanted by Dr. Serge Voronov, a Russian doctor who claimed the operation restored potency. Harold later married a not-terribly talented opera singer, Ganna Walska. His zealous promotion of her career is rumored to be part of the inspiration for *A Citizen Kane*.

Fowler did not fare much better than his father on the personal front, and was said to be troubled and complex. Fowler did not come to International until 1925, when a six-month stint working incognito for Harvester's Milwaukee foundry and encouragement from his wife and Alexander Legge convinced Fowler that his future was with International.

Harvester groomed executives in the field, and Fowler was no exception. He began as an apprentice working on experimental farm equipment and moved on to sales, where he was moved up to head of foreign sales. He then took over manufacturing operations and, in 1941, he was named president.

One of the largest challenges he faced was managing the company after World War II. This was one of the most properous times in company history, and also one of the most difficult.

Tractors sold at unprecedented rates in the late 1940s, but materials were in short supply due to the war. This created logistical challenges for IHC simply in terms of finding suppliers.

Under Fowler's leadership, management decided to expand. Harvester invested nearly $150 million in new factories, product designs, and machine tools. The company was reorganized, and new plants were built and purchased around the country to build everything from crawlers to Cubs to combines.

Fowler's critics would say the expansion was too much and the debt load acquired was too high for the profits that came in, but his personal reputation was that of a good man. John Sucher grew up in Hinsdale, Illinois, on a farm just down the road from the International Experimental Farm. His family was constantly using experimental tractors or watching International test tractors hours upon end in their fields and pastures. John recounts the day when Fowler McCormick came to visit their farm.

"He impressed me when I was just a kid. He pulled up to the farm in a big black limousine. The driver jumped out and ran over to open the door for Fowler, but he would have none of it. 'Forget it,' Fowler said. 'I can open the damn door myself!' He came right out in the cow yard to look over some equipment. There he was, in the cow shit with his oxfords. He didn't care. He was a blue-collar guy."

Fowler was chairman of the board from 1946 to 1951, and led the company through one of the most prosperous eras in the history of the company. Fowler McCormick died on January 6, 1972.

The Farmall H was the styled replacement for the F-20. It is also the best-selling single model of the 20th century, with more than 400,000 Model Hs and Super Hs sold.

appeared on steel during World War II, when rubber was reserved for war use. A lower first gear and taller fourth gear became available as optional equipment.

The clutch was a Rockford 10-inch single disk and was foot-operated. The brakes were external-contracting bands on forged steel drums mounted on differential shafts in housings. The brakes were foot-operated and could be engaged separately or in tandem.

Running Gear

The front axle and lower bolster pivoted as a unit. The original configuration of the Model H specified a narrow, dual-wheel tricycle front end, with the wide front end and single front wheel available as options. The standard front tread width was just less than 9 inches.

The rear end used the splined straight axles introduced on the F-12, allowing easy tread width adjustment by sliding the wheels on the splined axles. The rear wheels were reversible as well, allowing a broad range of rear tread width adjustment from 44 to 80 inches. The tractor weighed about 5,500 pounds.

The cotton duck-covered seat found on the smaller letter series tractors was standard on the H as well. The drawbar was an adjustable quick-attach model. The turning radius was 8 feet, 4 inches, and ground clearance was about 18 inches. Available options for the Model H included starting and lighting equipment, wheel weights, and rear wheel fenders. A temperature gauge and radiator shutter were standard with the distillate engine, and optional on Model H's with the high-compression gasoline engine. A rarer option was the wide-tread rear axle, which allowed adjustment from 44 to 100 inches.

Power Lifts

Three versions of Lift-All, International's hydraulic lift, were available for the Model H. The regular lift raised the entire attachment in one movement. A delayed lift was also available, which was used when both front and rear cultivators were mounted. The front implement lifted first, and the back unit a bit after, so the two effectively came out of the ground at the same point. The third Lift-All was the selective lift, which allowed raising left and right implements separately.

High-Compression

Gasoline Engine

In March of 1940, IHC authorized the addition of a high-compression gas-burning engine for the Model H. Despite the fact that the gasoline engine was not officially released until 1940, a gasoline-burning Model H was tested at Nebraska in 1939. The gasoline-burning Model H produced 19.14 horsepower at the drawbar and 24.28 horsepower at the belt pulley. One suspects that the tractors sent to Nebraska were pre-production machines, and that the publication of the gasoline tests led some to demand gasoline-burning versions for the showroom floor. A somewhat cryptic note on the decision

Farmall Model H, HV and Super H, HV Production	
Year	Production
1939	NA*
1940	41,317
1941	40,927
1942	34,987
1943	21,375
1944	37,265
1945	28,268
1946	25,615
1947	28,382
1948	32,265
1949	27,483
1950	24,681
1951	24,232
1952	16,243
1953	21,916†
1954	10,052
Total Prod.	415,008

†Super H introduced
*1939 shows about 10,000 serial numbers, but production schedule lists no figures. Models were definitely built in 1939, the production sheet may have just totaled them in the 1940 column for accounting purposes.
Source: McCormick/IHC Archives, "Tractor Production Schedule"

The high-clearance version of the Model H was the Farmall HV. A dropped-rear drive mechanism and a high-arch front bolster increased ground clearance for work on crops with high beds. This Farmall HV is preparing a seedbed for corn near Belle Rose, Louisiana, with a CIU-12 one-row chopper and a CIU-14 capper attachment.

to build the high-compression engine for the Model M lends some credence to this theory, although there is no way to be sure.

In August of 1940, the kerosene and distillate engine was modified to burn distillate fuel only. A kerosene attachment would be offered for those who wished to burn kerosene.

In December of 1940, the position of the controls were moved to the rear of the tractor for easier operation. A bracket was attached to the steering post, and the radiator shutter and magneto ignition switches were moved to the new bracket, as was the belt pulley control lever. The starting switch was moved to the platform, near the operator's foot, and the starting switch control rod was eliminated.

Farmall Model HV

In January 1942, a new model designed for cane and high-bed vegetable cultivation was released. The Farmall HV used step-down housings and an arched front axle to provide higher clearance. The front end had a whopping 30 1/4 inches of ground clearance.

The step-down housings provided the arch in the rear of the tractors. Inside the housings, a roller chain transmitted power from the drive sprocket to the rear axle. The housing was sealed, and the chain ran in an oil bath.

The dropped-rear drive used stub axles so that tread width was still adjustable. Additional adjustment was available by flipping the rear wheels, which had offset hubs. If the farmer required more adjustment, optional wheels with varying degrees of offset were available.

The engine, basic chassis, and running gear were the same as the Model H, and the tractor could be equipped with an assortment of special equipment.

Farmall Super H

Introduced in 1953, the Farmall Super H offered some of the new IHC line options, and the engine was punched out for a few more cubic inches and about 18 percent more horsepower. The H's bore of 3 3/8 inches was increased to 3 1/2 inches in the Super H, bumping displacement to 164 cubic inches and horsepower to 24 drawbar and 31 belt horsepower for the gasoline-burner. Torque was up to 267 foot-pounds at 1,046 rpm.

The cooling system was thermostatically controlled, oiling was done by pressurized lubrication, and the oil-bath air filter and replaceable oil filter kept things in the engine clean. The engine was advertised as being equipped with exhaust valve rotators, which were supposed to keep the valves rotating to prevent deposit build-up and extend valve life.

The Super H featured double-disk brakes rather than the drum brakes of the original Model H. The new brakes used a pair of rotating disks in a sealed housings. Lights and electric starter were standard, as was a PTO, belt pulley, Lift-All, and the new deluxe hydraulic seat.

Farmall Model M

The Model M chassis featured a more powerful engine fit in a chassis that used the same design as the Model H. The Model M was wider and heavier, and shared the same mounting points for implements. The Model H and M could interchange implements, but very few parts were shared on the actual tractors.

Chassis

As on the Model H, the Model M's rear frame, clutch housing, front frame channels, and upper bolster are bolted together to make a rigid

The Farmall M was the big tractor of the Letter Series. Rated as a three-bottom plow tractor, it shared the same chassis design as the Model H but had a larger, more powerful four-cylinder engine.

frame unit. The brakes (contracting bands on steel drums) and narrow front end were specified in the original decision on the Farmall Model M.

Engine

The Model M's four-cylinder I-head engine had a bore of 3 7/8 inches and a stroke of 5 1/4 inches. A variable governor was used, and the engine was rated for 36 flywheel horsepower at 1,450 rpm. The engine was started on gasoline, and then switched over to kerosene or distillate. The carburetor was an IHC model with a 1 1/4-inch updraft. Oil pressure was maintained with a gear circulating pump, and the oil filter used the Motor Improvement Company's P-20 star-shaped removable filter.

The engine was cooled with a closed system that circulated water through a flat-tube radiator with a water pump. The water temperature was originally supposed to be regulated with shutters in front of the radiator. In September of 1939, the company decided to use a belt-driven water pump and thermostat to cool the engine. This change was supposed to be effective on serial number 501, the first Model M off the line, so all Model Ms built should have had a thermostat. "Should" is the key word here, as decisions made in the executive boardroom didn't always make to the production floor in the desired time. The gas tank held 22 gallons and an IHC H-4 magneto with automatic impulse coupling and integral grounding switch was used.

Transmission

The transmission was a five-speed unit that used roller and ball bearings throughout with the exception of the reverse idler bushings. The tractor would run at over 16 miles per hour in top gear, which was intended for use only on pneumatic tires. The clutch was an 11-inch single-disk built by the Rockford Drilling Machine Company.

The rear tread width was wider on the Model M than on the Model H, with the M's tread width adjustable from 52 to 88 inches. A wide-tread model increased rear tread width to from 52 to 100 inches, although that model is quite rare. Ground clearance was the same on the two models at about 16 inches from the rear frame, and both shared the cotton duck-covered seat.

As tested at Nebraska in 1939, the distillate-burning Model M weighed 6,770 pounds (presumably including ballast) and delivered 24.89 horsepower at the drawbar and 34.16 at the belt pulley.

Gasoline-Burning Engine for the Model M

In February of 1940, IHC decided to supply the Model M with a higher-compression engine suitable for burning gasoline. The engine was otherwise the same, with the addition of a new cylinder head that brought the compression ratio up enough to burn 70- to 72-octane gasoline. The intake manifold was designed for gasoline operation, presumably without the fuel-warming system on the manifold for kerosene and distillate. This option was supposed to be available beginning with Model M serial number FBK-18144, meaning it was offered immediately after the decision was made early in 1940.

Interestingly, when the gasoline-burning version of the Model M was tested at Nebraska, it bore serial number FBK-ME533. Either the tractor tested at Nebraska used a prototype gasoline-burning setup, or the note on the decision to offer a gasoline tractor was incorrect. Either way, the gasoline-burning Model M produced a few more horsepower than the distillate machine. At the drawbar, the gas-powered Model M produced 26.23 horsepower with 36.07 horsepower at the belt pulley.

Tractor Wars, Round Two

Ford and IHC Square Off Again

Henry Ford seemed to have a penchant for causing troubles for IHC. His Fordson nearly put the company out of the tractor business entirely, and the Ford-Ferguson 9N took the shine off IHC's stunning debut of the Letter Series tractors. The Letter Series machines had it all: style, power, famed IHC reliability, and the latest bells and whistles. What they didn't have was the Ferguson System. Henry Ford got that, for a handshake no less, and was using it to jump back into the domestic tractor market.

The Ferguson System, or three-point hitch, was dreamed up by Harry Ferguson. It is ironic that it debuted on a Ford, because it solved the very problem that made widows of so many Fordson operators' wives. The Fordson was a bit light to plow well, so farmers would often connect the plow high up on the rear differential to force the plow into the ground. Great for hard soil, but if you hit a rock, look out. If the plow was stopped dead, the tractor would ratchet itself up and around. If the driver wasn't quick to either grab the clutch or leap off to the side, the tractor would pile drive the operator head-first into the ground.

What Harry Ferguson's system did was use a three-point mounting system that also forced the plow into the ground. In this case, though, the lever pushed down on the front wheels of the tractor. If you hit a rock, the front wheels would be pressed down on the ground. The system had some downfalls—erratic draft when plowing deep—but it was a tremendous improvement over what was on the market in 1939.

The concept was much like the one that led to creation of the Farmall: build the machine around the implement. The three-point hitch became an extension of the tractor. An ingenious system, and a patented one. Ford and Ferguson were the only ones who had it, and they were exploiting it to full measure.

Sales of the Ford 9N were substantial, with more than 30,000 sold per year from 1940 to 1942. The numbers were about one-third

The Ford 8N was another product that gave IHC fits on the sales floor. The 8N's big advance was the Ferguson three-point hitch, which offered superior draft control and plowing efficiency to other hitch systems on the market.

of IHC's sales of the Letter Series in those years, but about equivalent to sales of the Model H.

While IHC was looking for a hitch system of its own, Ford and Ferguson's famed "Handshake Agreement," in which they decided to join forces to produce Ford tractors with the Ferguson system, fell apart in 1947. Ford jettisoned Ferguson and launched a revised version, the 8N. Ferguson responded with a $340-million lawsuit and his own tractor, the Ferguson. Ferguson eventually won the lawsuit, and was awarded $10 million.

Ford and Ferguson's antics aside, the Model 8N began selling in big numbers. In 1948, Ford sold nearly 100,000 units. In 1949, sales of 8N topped that figure. While IHC was still selling significantly more total machines, no single IHC model approached sales of 100,000 in a year. Ford maintained a rate of sales close to that torrid pace until the mid-1950s, and did not leave the domestic market this time. While IHC remained the leader, the number of Fords pouring into the market had to bring up bad memories of how close the Fordson had come to driving IHC right out of the market in the 1920s

Later in 1940, the kerosene or distillate version of the Model M was changed to burn only distillate.

Farmall Model MD

In December of 1940, the Model MD was authorized. As the name suggests, the Model MD was simply a Model M with a diesel engine. The engine used the same bore and stroke as the gasoline and distillate Model M engines.

Starting was significantly different. The engine was started on gasoline with the engine in a low-compression mode. Once the tractor started and was warmed up, it was switched over to diesel fuel and high compression with a hand lever mounted on the rear fuel tank support.

The chassis was nearly identical to the regular Model M, with the possible exception of the brakes, which used gray iron drums rather than cast steel drums.

A hand-written note on the decision specifies that the first Model MD be serial number FDBK-26145. The date January 13, 1941, is also written on the decision. This was quite possibly the date the first Farmall MD left the factory.

Attachments for the Model MD included a 11-inch belt pulley, PTO, swinging drawbar, muffler, pneumatic tire pumps, wheel weights, spark arrestor, fenders, hydraulic power lift, dual rear pneumatic tires, and a low-speed option.

Farmall Model M Production

Farmall M, MV, MD, MDV

Year	Production
1939	6,739
1940	18,131
1941	25,617
1942	9,023
1943	7,413
1944	20,661
1945	27,479
1946	17,259
1947	28,785
1948	28,806
1949	33,065
1950	33,939
1951	43,405
1952	7,296
Total Prod.	297,718

Farmall Super M, MD, MV, MDV

1952	12,015
1952	1,905
1953	39,461
1953	10,636
1954	651
Total Prod.	64,668

Super M-TA, Super MD-TA

1954	23,523
Total Prod.	385,909

Source: McCormick/IHC Archives

The Super M was introduced in 1952. The bore of the engines was increased 1/8 inch, resulting in 264 cubic inches of displacement and increased power. The Super M also received disc brakes, standard electric starting and lighting, and more.

Farmall Model MV and MDV

In January of 1942, the company authorized the Farmall MV and MDV, high-clearance versions of the Model M. The Model MV used the distillate engine used on the Model M, while the Model MDV used the same diesel engine found in the Model MD.

A wide front end was used, with an I-beam forged axle mounted in the center to allow it to pivot. The axle was arched to provide additional ground clearance, and the front wheels were larger than those on a Model M. The rear was arched with a dropped-drive mechanism similar to the one used on the original Farmall. The rear tires were of similar size to the standard tires on the Model M. The final drive gearing was changed so that the five-speed transmission drove the MV and MDV at slower speeds. Top speed was about 15 miles per hour on the MV, and about 16 miles per hour on the MD.

Farmall Super M

By the 1950s, IHC was looking for ways to breathe some new life into the Letter Series. The line was by no means sagging, as 1950 was a record year for the Model M, but the farmer's appetite for power was slowly but surely increasing. The additions of hydraulics and an increasing number of PTO-driven implements and tools were cutting into the amount of power available to drive the wheels, and 35 horsepower didn't stretch as far in 1950 as it did ten years earlier.

The road to more power for the Model M took the predictable route; the Super M engine's bore was enlarged 1/8 inch to 4 inches while the stroke remained 5 1/4 inches. The result was a 264-cubic-inch engine that ran at a maximum of 1,450 rpm as the Model M. The engine was tested at Nebraska in 1952, and put out 33 drawbar horsepower and 44 horsepower on the belt pulley, an increase of over 8 horsepower. The gasoline engine produced 363 foot-pounds of torque at 991 rpm.

The diesel version was based on the same 264-cubic-inch engine. Like the gasoline engine, it produced peak horsepower at 1,450 rpm. The Super MD engine used a 12-volt system, and produced 350 foot-pounds of torque at 1,080 rpm. The diesel machine put out 33 drawbar horse-power and 47 belt pulley horsepower.

The Super M was also tested on LPG gas. The engine's compression was bumped up to 6.75:1 to burn LPG, and the result was more horse-power and torque. The Super M produced 371 foot-pounds of torque and 40 drawbar horsepower when burning LPG.

Note that IHC began noting displacement of its engines, going so far as to code them according to displacement. Also, the Nebraska Tractor Tests began measuring torque as well as horsepower. Torque is actually a better measure of pulling power, as it represents an engine's ability to increase rpm rather than the amount of power put out at a given rpm.

Horsepower Ratings

Tractors' horsepower ratings are often not that impressive when compared to the amount of power in vehicles of similar size. Considering that a decent V-8 engine in a 4,500-pound pickup truck puts out well over 200 horsepower, the Model M's 46 belt pulley horsepower in a 5,500-pound package sounds dreadfully under-powered. One factor is that the M's horsepower is measured differently, with the truck engine's figure typically taken at the crankshaft. Even so, the truck engine is putting out more than twice the horsepower of the tractor engine.

The Farmall Super M-TA was the most advanced Letter Series tractor built. The original six-volt system is often converted to a 12-volt system.

Gearing is a factor, but the real story is torque. Whether burning gasoline, diesel, or LPG, the Super M engine pumped out a torque peak of more than 350 foot-pounds at about 1,000 rpm. As any big-block muscle car enthusiast can attest, 300 or more foot-pounds of torque is serious power. The Super M was a powerful tractor in its day, capable of performing any task the average farm could dish out. The tractor still is relatively powerful. The obscene amounts of power put out by modern tractors overshadows the fact that the Super M is still a capable machine, and more than adequate for most daily tasks.

Super M-TA

The Super M-TA was a sign of things to come for IHC. It featured the company's latest innovations—Torque Amplification and an independent power take-off.

Torque Amplification (TA) was a high-low range that could be shifted on the fly. The drive was a planetary gear unit located between the engine clutch and transmission. Engaging it reduced tractor speed by one-third, giving the operator more pulling power to work with. The addition of TA was the beginning of increased flexibility in tractor transmissions, as manufacturers strove to give the farmer enough gear selec-

The Farmall Super MD-TA was the diesel version of the Super M-TA.

tions to match power output perfectly to load. The closer the match, the more quickly and efficiently the tractor could perform a task.

The independent power take-off (IPTO) was a simple concept that was complex to apply. The idea was to make the PTO engagement independent of clutch engagement. Previous PTO drives ran off the transmission, which was the most convenient place to find power in the rear of the tractor. An IPTO system was engaged entirely separately from the clutch, with a lever. To do this, power had to be transmitted directly to the rear of the tractor, bypassing the clutch. This was accomplished by driving the PTO directly from the flywheel on the Super M-TA.

The Super M-TA was produced in 1954 only, and was the forerunner to the Hundred Series. A diesel version was also available, and was known as the Super MD-TA or Super M-TA Diesel.

Chapter Eight

The Modular Economy

Styled Standard Tread Tractors

"The most beautiful curve is a rising sales graph."
—Raymond Loewy

The new-styled standard tread tractors were all about standardizing the line. The W-4 was the standard tread version of the Farmall H and the W-6 was the companion to the Farmall M. This practice cut costs in production and design, and also simplified things for the parts man at your IHC dealer. Instead of stocking four different sets of piston rings, only two were necessary. By the time you stock enough common replacement parts for a tractor, cutting the number of parts to stock by a third or more added up to significant space savings on the shelves. More importantly, the styled W tractors were built at Farmall Works, alongside the Letter Series tractors.

Standard treads were steadily but not wildly popular. The tractors were well-suited to plowing, industrial and orchard work. The low height worked well to sneak under orchard branches or through factories. Loewy styling and the latest features of the Letter Series Farmalls made the W-4, W-6, and W-9 solid additions to the IHC tractor line when they were introduced in 1940.

The gas and kerosene engines for the W-4 and W-6 were taken directly from the Farmall H and M. The WD-6 used the diesel engine from the Farmall MD, and the diesel and gas engines in the W-9 were from the TracTracTor line. The frame differed in that the H and M had front and rear channels that bolted to the block, while the W-4 and W-6 used front and rear gray iron frames bolted together. Mounting pads were cast into the frames, allowing attachments to be hung at several points. The distinction was fairly subtle, but the most important difference was that many of the parts were interchangeable.

Among the more interesting facets of the standard tread tractors were the names. As always, IHC was interchanging names to place on the tractors. The W tractors were first McCormick-Deering machines, then just McCormick machines with the new Loewy-designed "IH" logo on the side. Forty years after the merger, the names of the founding companies were still floating around. Actually, this probably had more to do with marketing strategies than anything else. Whatever grand scheme the company used to name the tractors, the orchard tractors fit into a similar plan to the W machines and bore McCormick-Deering early and McCormick with the IH logo later in production.

The industrial tractors—the I-4, I-6, and I-9—bore the International name, and had for most of the time. International was used more or less consistently with first industrial machines and, beginning with the hundred series, all of the standard tread tractors were Internationals.

It seems that not as many of the styled industrial tractors were intensively modified as with previous industrial models. The company advertising didn't push custom machines, and photos of heavily altered machines are scarce in the archives.

The standard tread tractors were the first IHC tractors to neatly merge wheeled and row crop production, a practice that became common. The tractors made lower production runs feasible and profitable, and were another example of IHC ingenuity bringing good equipment to the market.

The smaller of the so-called wheel tractors was the Four Series, which included the W-4, I-4, and O-4, all of which were introduced in 1940. Orchard tractors like this O-4 used a hand-operated throttle and clutch.

Four Series

The W-4 was the standard tread version of the Farmall H, of course, and used the same C-152 engine. The four-cylinder unit was started on gasoline and ran on kerosene or distillate. It used a variable-speed governor and was rated for 21 belt pulley horsepower at 1,650 rpm. As with the Farmall H, a high-compression gasoline-only version of the engine was released soon after the model appeared. The transmission was the same five-speed, with fifth a tall road gear that was locked out when used with steel wheels. The tractor was geared slightly lower than the Farmall H, with a 14-plus miles-per-hour top speed. The usual IHC filters for fuel, air, and oil were provided. The engine was cooled by a radiator and water pump system, but early models used shutters rather than a thermostat to control coolant temperature.

The front axle was a forged I-beam that pivoted on the front bolster. The steering mechanism differed from that of the Farmall H. On the Four Series tractors, it was a drag link that extended along the left-hand side of the tractor. Forged steel steering knuckles pivoted in bronze

Model O-4, OS-4, W-4, I-4, and Supers Production	
Year	Production
1940	213
1941	2,741
1942	1,888
1943	1,759
1944	3,705
1945	2,799
1946	1,893
1947	2,822
1948	2,975
1949	2,973
1950	3,186
1951	3,519
1952	2,385
1953	2,566
1954	444
Total Prod.	35,868

Source: McCormick/IHC Archives; Tractor Production Sheet

bushings in the ends of the front axle. The rear axles were mounted in removable housings. The tractors were reasonably light, at about 3,800 pounds with single rear wheels, and about 5,500 pounds with dual rear wheels. The heavy-duty option, which included beefier front and rear axles and hydraulic brakes, added about 600 pounds more.

The basic standard tread, the W-4, was the first model that appeared. It was approved for production in March 1940. The orchard version, the O-4, was approved shortly thereafter, in April 1940. The O-4 was the same machine with the addition of smooth body work and swoopy fenders that prevented damage to trees as the tractor slipped through the orchard. An underslung muffler and air intake and a hand clutch completed the package, keeping the machine low and compact. Orchard models also have the distinct advantage of stylish Buck Rogersish good looks.

The other Four Series was the International I-4, the industrial version of the W-4. The sturdy I-4 could be equipped with dual rear tires (which were, as you probably know, actually quad rear tires, but never mind), rear work and travel lamps, front and rear wheel weights, and hydraulic remote control.

In 1953, the Four Series went Super, and each tractor was labeled with the "Super" on the decal on the front of the hood. Under the hood, the new C-164 engine was up to 164 cubic inches by increasing the bore 3/8 inch. Horsepower took a corresponding jump, up to about 30 horsepower at the belt when burning distillate and 33 on gasoline.

Six Series

The first of the Six Series machines to appear was the W-6, which was the wheeled counterpart to the Farmall Model M. Just as the Model

H and M shared a common chassis design, the Four and Six Series machines were much alike. The significant difference was the engines, especially the diesel engine available in the Six Series. The increased power had the tractor rated to pull a three- to four-bottom plow or an 18-foot single disk.

Engine

The base model was the W-6, which was authorized for production in November 1939. Production records show that it hit dealership floors in 1940. The tractor used the distillate- or kerosene-burning engine from the Model M, which was the 3 7/8 x 5 1/4-inch four-cylinder. A variable governor was used and the engine was rated for 31 horsepower at the belt and 24 horsepower on the drawbar when burning distillate fuel. The original cooling system used a radiator and water pump with shutters to regulate engine temperature, although later models used a thermostat. As with the Model M, the high-compression gasoline engine was offered as an option in mid-1940 for all Six Series tractors.

Interestingly, in July 1940 IHC decided to authorize offering a kerosene-burning attachment for the standard Six Series and Farmall M tractors, despite earlier models being clearly marked as capable of burning distillate or kerosene. It's possible the earlier engines did not burn kerosene well, and the kerosene attachment addressed that problem. All standard Six Series tractors were rated to burn distillate. To burn kerosene, at least according to IHC, you had to purchase the kerosene-burning attachment.

The chassis and running gear were similar to those of the Four Series. The transmission was a five-speed, with a tall fifth gear locked out on steel-wheeled tractors. The clutch was an 11-inch Rockford Drilling Company single-plate unit. The brakes were external contracting bands on drums. The Six Series tractors weighed in at about 4,800 pounds with single rear wheels, about 1,000 pounds heavier than the Four Series tractors.

WD-6

The first variant to be introduced was the WD-6, the diesel version. Authorized in February 1940, it should have been on the farm by the harvest season of that year, and in significant numbers by Spring 1941. The D-248 engine was also used in the company's small crawler, the TD-6. The four-cylinder engine used the same bore and stroke as the

The Six Series were the wheel tractor versions of the Farmall Model M. Early Four and Six Series tractors used the typical radiator and water pump cooling but used shutters to regulate temperature. Later versions used a thermostat.

kerosene/distillate engine, and was started on gasoline. Despite the similar displacement, the diesel put out a few more horsepower than the standard W-6 engine, with the WD-6 pumping out 33 horsepower on the belt and 26 horsepower at the drawbar.

The chassis, clutch, and transmission were identical to the W-6. The cost was slightly higher, but the increased efficiency and horsepower of the diesel engine should have made it a popular choice with farmers. No differentiation was made in serial numbers, so production figures are not available.

O-6

The next model to appear was the O-6, the orchard version. As with the O-4, the keys to an orchard model were streamlined fenders and underslung exhaust and intake pipes. The kerosene/distillate engine was used on the O-6, while the OD-6 used the C-248 diesel engine. The O-6 was authorized for production in mid-December 1940, making it available early in 1941. The other Six Series machines were the industrial versions, the I-6 and ID-6, which were introduced in 1940.

Supers

The Six Series went Super in 1952, with engines providing increased power. The horsepower gains were found with an increased bore, up to 4 inches, with the same 5 1/4-inch stroke. The standard engine became the C-264 gasoline burner, and horsepower was up 8 horsepower at the belt to 41 and increased 7 horsepower on the drawbar to 33. Bear in mind that the increases were magnified by the fact that the new engine burned gasoline, which produced 5–10 percent more horsepower than distillate.

The D-264 diesel engine used the same bore and stroke as the gas engine, and put out the same amount of horsepower. The diesel equipment was a few hundred pounds heavier, but otherwise the Super W-6 and Super WD-6 were nearly identical.

The last variant of the Six Series were the TA versions, the Super W6-TA and Super WD6-TA. Like the Super M-TA, these two tractors incorporated torque amplification and live PTO. Produced only in 1954, these were a sign of things to come for IHC. Only about 3,000 or fewer of these tractors were built, making them extremely collectible.

Model O-6, OS-6, ODS-6, W-6, WD-6, I-6 and ID-6 Production	
Year	Production
1940	287
1941	2,558
1942	1,513
1943	1,180
1944	2,584
1945	4,852
1946	3,273
1947	5,322
1948	5,251
1949	5,498
1950	4,735
1951	5,875
1952	3,838
1953	6,627
1954	3,089
Total	56,482

Source: McCormick/IHC Archives, "Tractor Production Schedule"

Nine Series

The Nine Series brought additional power to the farmer, and offered the farmer the simple improvement of increased capabilities. While the Four and Six Series machines were significant by integrating IHC lines, the Nine Series models' importance stems from their individuality. The Nine Series were big, open-field tractors that used powerful engines from the crawler series, and had an integral-frame chassis design to meet the demand for more powerful tractors. It's no coincidence that these machines were the best-selling of the wheel tractors; the farmer was demanding more and more power on the farm. A combination of an increase in driven implements, larger implements, and changing tastes made horsepower the key to selling tractors, especially in the 1950s and 1960s. The Nine Series was IHC's first entrant into the horsepower race.

W-9

The first of the Nine Series, the W-9 was introduced in 1940. Like the Nine Series precursor, the W-40, the W-9 used the same engine as the T-9 crawler, the C-335. That engine could be equipped to burn gasoline, kerosene, or distillate, and pumped out 44 horsepower at the belt and 36 horsepower at the drawbar. The diesel engine was actually quite similar, with identical bore and stroke (4.4x5.5 inches) and similar horsepower output. Diesel Nine Series were indicated with a "D" in the model name (WD-9, ID-9, etc.).

All used the same chassis and running gear. The transmission was a five-speed, with a 12-inch single-plate clutch. As with the rest of the IHC line, fifth gear was locked out on tractors with steel wheels. The magneto was an IHC H-4, and the carburetor was also an IHC unit, a Model E-13. The cooling system used a water pump and radiator and shutters initially, with later models using a thermostat without the shutters.

Optional equipment for the Nine Series machines included a PTO, swinging drawbar, pneumatic tire pumps, wheel weights, and a wide assortment of steel wheels and rims and pneumatic tires.

A Nine Series tractor was the largest, heaviest machine in the IHC line, weighing in at about 10,000 pounds with fluids and weights. The weight and horse-

Nine Series Production

I-9, ID-9, W-9, WD-9, WR-9, WDR-9, and Supers

Year	Production
1940	0*
1941	1,764
1942	1,263
1943	1,220
1944	5,765
1945	5,868
1946	5,373
1947	6,235
1948	6,247
1949	9,485
1950	6,937
1951	7,231
1952	5,815
1953	4,773**
1954	3,662
Total Prod.	74,141

*IHC documents conflict on 1940 production; serial numbers are listed for that year, but the production chart shows no Nine Series tractors built in 1940.
**Super Nine Series production began in 1953
Source: McCormick/IHC Archives

The W-9 was the largest International wheel tractor, and the Nine Series appeared in 1940. Unlike the Four and Six Series, the Nine Series chassis was unique rather than based on another model.

power translated into sheer pulling power and traction in any conditions.

WD-9, WR-9, and WDR-9

Three variants of the Nine Series were built, the regular W-9 and WD-9, the rice field special WR-9 and WDR-9, and the industrial I-9 and ID-9. Perhaps the most unusual of these were the Rice Field Specials. Rice fields are moist and loamy, and require tractors with high horsepower and good traction. The Nine Series were well-suited to the task, and the original Rice Field Special was equipped with steel wheels with large spades for traction. Ensuing models used wide pneumatic tires with deep lugs, which must have worked much better than steel.

Super Nine

The Super Nine tractors used an increased bore to bring displacement up to 350 cubic inches. The four-cylinder engine had giant pistons, with a 4 1/2-inch bore and 5 1/2-inch stroke. The engine could be equipped to burn gasoline, diesel, kerosene, or distillate. The pistons were aluminum, and the cylinder sleeves were replaceable. The diesel engine used fuel injection and had a compression ratio of 15.6:1. The powerful engine put out 530 foot-pounds of torque at 1,039 rpm. The Super Nines were produced in 1953. Essentially the same tractor was produced until 1958 as the 600 and the 650, which were simply Super Nines with slightly different standard and optional equipment.

The Nine Series demonstrated IHC's vision. It was the most successful of the wheel tractors, and was what the farmer of the day wanted and needed; more horsepower. Although horsepower was not yet the deciding factor in a purchase, it would be one of the most important considerations from the late 1950s to modern times.

Chapter Nine

Expansion and Diversification

Styled TracTracTors

"Beauty through function and simplification."
—Raymondy Loewy

When Raymond Loewy was styling International products, the TracTracTors received the same smooth, functional styling as the Letter Series tractors. The first to be restyled was the big fuel-injected TD-18, with prototypes floating around as early as 1938. The next to be styled were the new T-6, T-9, and T-14. The two smaller crawlers were part of IHC's efforts at integrated production, as the T-6 shared componentry with the Model M and W-6, and the T-9 shared parts with the W-9. A few of the machines were introduced in 1939, and the full line was put into production in 1940.

All of the new crawlers featured replaceable cylinders and main bearings, with the cylinders constructed of heat-treated sleeves machined to fit on the outside and honed and polished inside. The engines ran on

The T-6 and TD-6 were smaller machines, weighing about 7,000 pounds and putting out about 30 drawbar horsepower. The original T-6 and TD-6 used the same engines as the W-6 and WD-6.

roller and ball bearings throughout, and were cooled with a thermostatically controlled radiator and water pump system. Steering clutches and brakes were used for steering, as with previous TracTracTors. Most of the models were available in a narrow- and wide-tread versions.

Available options included cabs, electric lights and starting, a variety of track lugs (often called "grousers"), cutaway drive sprockets for heavy mud, radiator shutters for cold weather, and an assortment of guards, bumpers, and shields.

T-6 and TD-6

The T-6 and TD-6 were the crawler counterparts to the Model M and Six Series. They used the same engines, but the chassis was quite different. The crawler chassis was heavier than the tractors, weighing about 7,000 pounds dry.

The T-6 used the gasoline, kerosene, or distillate engine, while the TD-6 used a diesel unit. Both had a five-speed forward and one-speed reverse transmission, with the top forward speed a blistering 5.4 miles per hour. They were available as both wide- and narrow-tread machines, with a 40-inch tread for the narrow model and 50-inch for the wide tread model. When tested in 1940, the gasoline-burning T-6 produced 33 belt and 25 drawbar horsepower on gasoline and 31 belt and 24 drawbar horsepower on distillate. The TD-6 produced 31 belt and 22 drawbar horsepower when burning fuel oil. It was rated to pull a two- to four-bottom plow.

The T-6 and TD-6 received larger engines in several later variations, known as the Series 61 and Series 62. The small crawlers were replaced with the T-4, T-5, and TD-5.

T-9 and TD-9

The T-9 and TD-9 were the counterparts to the Nine Series. The crawlers weighed about 10,000 pounds, and were available in narrow (44-inch) and wide (60-inch) tread widths. Like the smaller crawlers, five-speed forward and one-speed reverse transmissions were used, with a top forward speed of 5.3 miles per hour.

The T-9 gasoline engine produced for 41 horsepower at the belt, 32 at the drawbar when burning gasoline. Rated to pull a three- to five-bottom plow. The T-9 used the same C-335 engine as the Nine Series wheel tractors. The T-9 was produced only a short time, from 1939 to 1941, probably due to the popularity of the diesel engine.

The TD-9 engine was rated for 39 horsepower at the belt, 29 horsepower at the drawbar when introduced. It used the D-335 engine, the same that was used in Nine Series diesels. The TD-9 was produced from 1940 to 1956. During that time, the engine's power output was slightly increased. When tested at Nebraska in 1951, power was up a couple of horsepower at the drawbar and the belt.

In 1956, the Series 91 was introduced. The engine bore was increased to bring the four-cylinder diesel's displacement up to 350 cubic inches. The compression ratio was 16:1, and horsepower was up to 43 horsepower at the drawbar and 55 at the belt when tested at Nebraska in 1956.

In 1959, the Series 92 TD-9 came out with a new turbocharged six-cylinder 282-cubic-inch engine. The new engine used a 18:1 compression ratio and brought horsepower up to 66 at the PTO and 44 at the drawbar. The Series 92 was replaced by the Series B in 1962, and production continued through at least 1973. At some point, the TD-9's name was changed to the Model 150 crawler, probably around 1969 when the new Model 100 and 125 crawlers were introduced.

The TD-14 had a six forward- and two reverse-speed transmission and a top speed of nearly 6 miles per hour. A four-cylinder 461-cubic-inch diesel powered the crawler. The TD-14A was produced later in the production run.

TD-14

The original TD-14 was produced from 1939 to 1949. The tractor was significantly larger than the TD-9, weighing about nine tons with weights and fluids. The TD-14 was available in a wide or narrow tread, with either a 74- or 56-inch tread width.

The original version of the TD-14 was produced from 1939 to 1949. The four-cylinder 461-cubic-inch diesel engine produced 52 drawbar and 62 horsepower at the belt pulley when tested at Nebraska in 1940. The transmission had six forward and two reverse speeds, with top forward speed at 5.8 miles per hour.

The TD-14A was produced from 1949 to 1955. The Series 141 TD-14 was produced from 1955 to 1956. The engine size remained the same 461 cubic inches. The Series 142 was introduced in 1956, and production ran until 1958. It also used the same-size engine, but power output was up due to an increased rpm rating. The TD-14 put out 61 drawbar and 83 belt horsepower when tested at Nebraska in 1956.

TD-15

The TD-15 was the replacement for the TD-14. Introduced in 1958, production ran until at least 1973. The TD-15 featured a new 554-cubic-inch six-cylinder diesel engine. Tested at Nebraska in 1960, the new engine brought the TD-15 up to 77 drawbar horsepower (PTO horsepower was not tested). The increase was not that substantial, considering the engine was much larger than the older TD-14 powerplant. The tractor's weight with weights and fluids was up substantially, at more than 12 tons. A 12-volt electrical system was used.

TD-18

The development of the TD-18 began in the mid-1930s, when the company experimented with a large crawler. The TD-18 was larger than any tractor the company had produced previously. It was built in several versions from 1939 to 1957, with each iteration producing more horsepower. The engine was a six-cylinder diesel rated for 85 horsepower on the belt and 70 drawbar horsepower (later versions had 109 belt horsepower). The tractor was huge compared to anything else the company built, weighing in at over 10 tons. The tracks were 84 5/8 inches long and 18 inches wide, which put 3,046 square inches of track on the ground.

The six-cylinder diesel engine had a bore and stroke of 4 3/4 x 6 1/2 inches. The engine featured Bosch fuel injection, full-pressure lubrication, and removable cylinder sleeves and main bearings. The engine was cooled with a thermostatically regulated radiator, water pump, and fan system, with optional shutters for cold weather.

TD-24

The TD-24 was International's grand entrance into the construction business, and it debuted a host of new features in an 18-ton package touted as the most powerful in the industry. The TD-24 suffered some problems when it was introduced, but carried the construction banner for IHC capably for more than a decade.

The TD-24 was produced from 1947 to 1959. The power output was stunning for the era, and is still substantial today. International advertised that the TD-24 was the most powerful crawler tractor in the industry. When tested at Nebraska in 1950, the tractor was too powerful for the dynamometer so belt horsepower could not be tested. The tractor recorded a stunning 138 drawbar horsepower. Power output rose steadily as the model was developed, and the 1954 edition cranked out 155 drawbar horsepower at the Nebraska text, and the 1957 machine was up to 163 horsepower at the drawbar.

The crawler weighed about 20 tons, and was powered by a 1,090-cubic-inch six-cylinder diesel that carried 28 quarts of oil in its sump. The tractor used an unusual 24-volt electrical system and a 317-pound crankshaft.

Planetary Power Steering

One of the new features on the TD-24 was what International called Planetary Power Steering. This feature combined hydraulically controlled steering clutches with a two-speed range similar in function to the Torque Amplifier introduced on the Super M-TA. The dual-range unit could be engaged on the fly, allowing shifting under power. The two-speed unit was coupled to a four-speed synchromesh transmission, giving eight speeds in forward or reverse. The synchromesh transmission was nearly unprecedented in the 1950s, and International tractors did not receive synchromesh transmissions until the 1980s.

The two-speed system could be utilized for gradual turns by dropping one track to the "low" position and the other to "high." This provided power to both tracks while turning. To pivot, the inside track was stopped and only the outside driven, which was the typical method of turning a crawler.

Troubles with the TD-24

Despite all these advances, the TD-24 came out with some flaws and, according to Barbara Marsh in *A Corporate Tragedy*, was an embarrassment for IHC. The TD-24 apparently broke down quickly, and Marsh contends that the TD-24 problems badly tarnished IHC's reputation in the heavy equipment field, damaging the company's attempts to compete with Caterpillar and the big Allis-Chalmers crawlers.

The TD-24 was hardly the last of the TracTracTor line, as IHC crawlers continued to flow off factory floors in a fairly steady stream. It is the end of the crawler story in this book, tragically. The line was headed a bit south after the TD-24, and times got increasingly tough for the entire agricultural equipment industry. At one point, IHC was selling crawlers at a 10 percent loss, banking on service parts to make up the difference. The strategy may have worked, but that's a hard way to make money. The tough times took their toll, and the late 1960s and early 1970s saw mergers become the rule of the day for tractor manufacturers. Where once there had been over 180 tractor manufacturers in the United States, there would be only a handful left by 1970. A key to keeping IHC afloat was the crawler line, although their positive impact on profitability lessened after the ill-fated TD-24.

TD-25

The TD-25 was introduced in 1959, and was the replacement for the TD-24. The big TD-25 used a turbocharged 817-cubic-inch six-cylinder diesel to crank out a prodigious 187 drawbar horsepower. The crawler weighed 26 tons and used a synchromesh eight-speed transmission. A 24-volt electrical system was used, and the crawler was produced until at least 1973.

TD-30

The TD-30 was another monstrous crawler designed for the construction market. It was introduced in 1962, and production ran until 1967. The TD-30 used the same basic engine as the TD-25, although the new crawlers were larger, at about 29 tons. The TD-30 was available with either a power-shift or gear-drive system.

The TD-30 six-cylinder engine in the power-shift model was both turbocharged and intercooled, and was rated for 435 horsepower. The same six was used in the gear-drive model, but without the intercooler, resulting in a rating of 375 horsepower. Both engines displaced 817 cubic inches and used fuel injection, four-valve heads, and a massive oil cooler.

The power-shift drive was similar to the drive used on the TD-24, and was a torque converter coupled to several hydraulically-controlled clutch packs. The result was similar to that of an automatic transmission, with four simple speeds and the dual-range Planetary Power units controlling forward and reverse motion. The power-shift model used a hydraulic fluid cooler, and had more horsepower than the gear-shift model. The torque drive ate up the extra power, making the two models comparable in drawbar performance.

A synchronized gear transmission was also offered. The transmission was a four-speed unit, and the addition of the Planetary Power units from the TD-24 gave eight speeds in forward and reverse. A huge 16-inch dry-disk clutch transmitted power from the engine to the transmission, and the synchronized transmission allowed gears to be shifted while the crawler was moving.

A complete line of construction attachments was offered, including dozer blades, rippers, push plates, and all sorts of guards and cages.

T-340 and TD-340

Produced from 1959 to 1965, the 340 crawlers used the engines from the diesel and gas 340 tractors. The crawlers weighed about 3 1/2 tons with fluids and weights. Like the earlier crawlers, the 340 crawlers were available in wide- or narrow-tread configurations, with tread widths of either 38 or 48 inches.

The 135-cubic-inch gasoline engine in the T-340 was rated for 35 belt and 31 drawbar horsepower. The compression ratio was 8:1, and early models used a 6-volt ignition (a 12-volt was used by 1963). The four-cylinder engine could be run to 2,000 rpm.

The 166-cubic-inch diesel engine in the TD-340 tested at 32 drawbar and 36 PTO horsepower at Nebraska in 1960. The larger diesel was also a four-cylinder, but bore and stroke were completely different. The compression ratio was 19:1, the higher ratio necessary to burn diesel.

The transmissions were five-forward and one-reverse-speed units. The 340 crawlers could be equipped with either a Torque Amplifier or a Fast Reverser. The Torque Amplifier allowed on-the-go shifting between a high and low range. The Fast Reverser gave an instant reverse gear that was 22 1/2-percent faster than the forward speed selected. Both options could not be mounted on the same machine.

Case still makes a line of construction equipment, including crawlers and skid steer loaders.

Regular equipment included double-disc brakes, a cigarette lighter, tachometer, heat and fuel gauges, and a swinging drawbar. Options included a three-point hitch, "live" hydraulic system, independent 540-rpm PTO, lights, radiator shutters, and an assortment of guards.

T-4, T-5, and TD-5

The T-4, T-5, and TD-5 were the smallest crawlers in the line, and were introduced in 1959 or 1960. The tractors used a derivative of the engine from the Super C in a 3 1/2-ton frame. These crawlers weight just as much as the more powerful 340 crawlers. Weight is often an asset in a crawler, as it increases traction and drawbar pull in most cases.

Late Crawler Production

Model	Years	Est. Production*
T-4	NA	NA
T-5, TC-5	NA	NA
TD-5	NA	NA
T-6 & TD-6	1940–56	38,550
T-6 & TD-6 (61 Series)	1956–59	2,500
T-6 & TD-6 (62 Series)	1959–69	3,920
Total T-6 & TD-6	**1940–69**	**44,970**
T-7	1969–73	1,040
T-8	1969–73	1,030
T-9	1940–43	10,920
TD-9	1940–56	59,800
TD-9 (91 Series)	1956–59	6,990
TD-9 (92 Series)	1959–62	6,000
TD-9 (Series B)/150	1962–73	10,030
Total T-9 and TD-9	**1940–73**	**93,740**
TD-14	1939–49	26,260
TD-14A	1949–55	12,540
TD-14A (141 Series)	1955–56	2,250
TD-14 (142 Series)	1956–58	4,010
Total TD-14	**1939–58**	**45,060**
TD-15 (150 Series)	1958–61	3,500
TD-15 (151 Series)	1961–62	680
TD-15	1963–73	21,150
Total TD-15	**1958–73**	**25,330**
TD-18	1939–49	22,040
TD-18A	1949–55	11,110
TD-18A (181 Series)	1955–56	2,450
TD-18 (182 Series)	1956–58	2,760
Total TD-18	**1939–58**	**38,360**
TD-20 (200 Series)	1958–60	2,500
TD-20 (201 Series)	1961–62	470
TD-20	1963–73	23,000
Total TD-20	**1958–73**	**25,970**
TD-24	1947–55	7,500
TD-24 (241 Series)	1955–59	3,630
Total TD-24	**1947–59**	**11,130**

TD-25 (250 Series)...................... 1959–62.. 1,290		
TD-25 (Series B) 1962–71 3,000		
TD-25................................. 1968–73... 1,900		
Total TD-25 ....1959–73..6,190		
TD-30................................. 1962–67... 690		
100 .... 1969–73... 690		
T-250..................................... 1963–73..6,440		
125 .... 1969–73.. 1,000		
T-340 and TD-340....................1959–65 8,030		

*Production figures are number of serial numbers and do not include the last year of production. These figures are only rough approximates and are rounded off.
Source: Crawler Serial Line Numbers, IHC, 1973. NUMBERS ARE ONLY VALID TO 1973.

The smaller T-4, T-5, and TD-5 do lack the more sophisticated hydraulic system available on the 340s, but the smaller machines compared favorably to the competition. John Deere, Case, and Oliver all had crawlers of a similar size, and the International T-4, T-5, and TD-5 have more power and more or equal pulling power. They also had more standard transmission speeds.

T-4

The T-4 used the same C-123 engine as the Super A, rated to run at a maximum of 2,000 rpm. The T-4 put out 21 drawbar and 27 PTO horsepower when tested at Nebraska in 1960. The crawler was equipped with a Torque Amplifier.

T-5

The T-5 used the 135-cubic-inch engine, which looks to be a longer-stroke version of the Super A engine. The slightly larger engine revved to the same 2,000 rpm, and produced a little more horsepower, generating 29 at the drawbar and 36 on the PTO. The T-5 was tested in 1960 and was equipped with a five-speed transmission and a Torque Amplifier, like the T-4.

TD-5

The TD-5 used a 144-cubic-inch diesel engine from the B-275 tractor. The fuel-injected four-cylinder diesel used a 21:1 compression ratio, and produced 30 drawbar and 35 PTO horsepower. The TD-5 tested in 1960

Tracks live on in the Case-IH line.

was equipped with a five-speed transmission with a Torque Amplifier, just like the other two small crawlers.

The crawlers were available in three tread widths of 38, 48, and 68 inches. The tractors had a standard five-forward and one-reverse-speed transmission that could be equipped with a Torque Amplifier. Two versions of live PTO could be ordered, the standard type and what they called the constant-running PTO. A two-position clutch allowed the operator to depress the clutch part way and disengage the wheels but not the PTO, while depressing the clutch all the way disengaged the engine and the PTO. International called this two-position system a constant-running PTO.

Several different hitch and hydraulic options were offered. A three-point hitch was available with an assortment of hydraulic systems to control just the hitch or a remote implement and the hitch. A straight swinging drawbar was offered as well. A four- or five-roller frame and an assortment of track shoes were also offered.

Chapter Ten

The Tractor in the Garden

Cub and Cub Cadet

"Suburbia is where the developer bulldozes out the trees, then names the streets after them."
—Bill Vaughn

The Farmall A, B, and C were designed to fit the needs of the smaller farmer who was perhaps making the change from horse farming to tractors, and the Model H and M satisfied the typical farmer. The Cub and Cadet were targeted at an entirely different group and were in some ways more important to International's future than the more traditional tractors.

Opening new markets was crucial for IHC, especially in the post-war period. For one thing, the company had geared up production during World War II, building crawlers, bombs and more for the government. With that demand suddenly gone, the company was left with a large work force that would have to be scaled back unless the company expanded. The company chose expansion, attacking new markets like milk coolers, construction equipment, and even some appliances such as refrigerators and air conditioners.

The company also looked to expand coverage in existing markets. One of the trends that had to concern tractor manufacturers was the fact that the number of farms without tractors was dropping off. In the 1920s and 1930s, horse use was still prevalent, and manufacturers had to figure out how to build a tractor that would replace the horse. The market was almost limitless, and sales potential was huge.

The sales potential of the post-war era was even larger, as the tractors of the 1940s were enough of an advance that most farmers found that replacing the tractors of the 1920s or 1930s with new machinery was a profitable move. Also, there were still a respectable percentage that hadn't turned to power farming. An incredible 1.5 million tractors were added to farms in the United States between 1940 and 1950, but that trend simply could not continue. Once most of the farms had a modern tractor, the sales volume would drop.

Savvy tractor manufacturers, like International, were aware of this. The only way to keep moving machinery was to develop radical new technology or cater to farming niches. The bigger tractors entered the horsepower race and received new developments, but the key would be selling tractors to people who didn't have them yet. So the tractor line was broadened to include more special equipment for rice farming, cotton growing, sugar beet producers, and orchard farmers. New plants were built just to cater to these types of markets, and new equipment was designed that performed these tasks more efficiently than ever before.

While searching for farming niches, International must have discovered the fact that more than three million farms still didn't have tractors. Small farms of less than 50 acres did not typically have a tractor. Some of these farms were vegetable, tobacco, and livestock farms. Many of these farmers were part-timers and had another source of income. This huge untapped market was the target for the Cub.

Some say that the Cub was built for tobacco farmers. Although the Cub was indeed built for small farms, and tobacco farms were typically about 25-acre operations, the tobacco farm was only a small portion of the market. The Cub was designed to be small and cheap enough to lure new buyers who perhaps dreamed of a tractor for their little operation, but found previous models was too big and expensive to be practical.

The Cub Cadet was a direct outgrowth of this approach. Once the smaller farmer had a tractor, who was left? The home owner, of course, and most of the residents of the United States were in this category. Better yet, the Cub Cadets were the type of vehicle that gets replaced every few years. The Cub Cadet was the beginning of America's move to lawn tractors, a machine that is hard to imagine living without today. But it all began with the Cub.

Farmall Cub

The Cub was the left hook of IHC's one-two combination for the late 1940s. The little tractor was a great cultivator, could pull small implements, and was dirt cheap to purchase and operate.

The Cub was seen as early as October of 1945, when it was one of the star attractions at a show at International's experimental farm in Hinsdale, Illinois. The Cub was promoted heavily in 1946, and was introduced as a 1947 model. It was produced as the Farmall Cub until 1965, when the Farmall name was dropped in favor of International. The International Cub was built in various forms until 1979.

A New Machine

From the name to the introduction hoopla, the Cub was aimed squarely at a new and different market. The rest of the line bore letters or more utilitarian names. The Model M just sounds big and powerful, and later models had especially manly names. "Super M-TA" just has a ring to it that tells you this is a heavy piece of equipment. Not to mention the Titans, Moguls, and even the Farmall had names that connoted work and power. This was not so with the name, "Cub."

Early Letter Series advertisements referred to the larger machines as the "bears." With this in mind, one can see where the Cub name originated. With the first letter of the alphabet already assigned to a small tractor, the company could not go smaller alphabetically. But "Cub" was a natural for the progeny of the "bears" and fit with the concept that the Cub would be marketed to a group of people who previously considered tractors too large and expensive for their farms.

The Cub was introduced with trumpets blaring and flags waving. Cubs appeared in parades. Cubs were put on display in banks and at

The Farmall Cub was introduced in the fall of 1945, and promoted heavily across the country. This woman is trying a new Cub out at a huge showing of International tractors and equipment held at IHC's Hinsdale Experimental Farm in October of 1945.

fairs. Several promotions piled herds of Cub Scouts on the little tractors. Promotional banners and displays proclaimed ease of use and affordability. The archival photographs show a disproportionate number of women operating Cubs. Are you starting to see a pattern here?

All the signs indicate that IHC's management—a very bright group— knew exactly what they were doing with the Cub. They knew that there was a huge, nearly unlimited market of folks out there without tractors who could surely use one to haul garbage, till the garden, maybe even cut the grass. Owners of large country homes and smaller farms were the starting point, obviously, and the next stop was suburbia. And the way to get into these smaller places was through the person who really controlled the money spent in the household—the woman of the house. Early market research showed that—at least in the automotive industry—buying decisions were the domain of woman of the household, despite the fact that the man was more likely to write the check.

So, the newest tractor from IHC had a friendly name and was promoted with kids, in banks, at parades, and anywhere else the general public (rather than just farmers) gathered. The company was attempting to bring the tractor to a new group of people, and management believed the company could do so in large numbers. In 1947, a corporate brochure celebrating the 100th year anniversary of the invention of the reaper touted the Cub as a rising star. Production was expected to top 50,000 per year, and Louisville Works, where the Cub was built, was gearing up to meet that anticipated demand.

International tracked the first 10,000 Cubs sold very closely and came up with some fascinating statistics. Over half of the Cubs went to vegetable, grain, and livestock farmers, with only 3 percent going to tobacco farmers. The tobacco myth was shattered by this statistic. If the company had actually built the Cub with tobacco farmers in mind, the sales to them were a dismal failure. If the tractor was intended for small farms, the tractor was a smashing success. About 75 percent of the Cubs went to farms of less than 50 acres. The Cub tended to be the only tractor owned, with 67 percent of the buyers not owning any other tractor. Also, nearly half of the Cub owners weren't even full-time farmers.

International knew it was backing a winner with the Cub, and did everything possible to see it succeed. The phenomenal success of the Cub was no fluke.

Cub Nuts and Bolts

The Cub used the same integral frame construction as the Model A and B, with the engine, transmission, clutch, and rear differential housings serving as the frame. The rear tread width was adjustable from 40 to 56 inches, and the relatively tall, wide front end and arched rear differential gave generous ground clearance. Like its fellow Farmalls, the Cub was well-suited to cultivating, and could pull a single 12-inch plow with relative ease.

Engine

The engine was an IHC 60-cubic-inch four-cylinder L-head with a bore of 2 5/8 inches and a stroke of 2 3/4 inches. The relatively short-stroke engine was rated to run at a slightly higher rpm—1,600—than the larger, longer-stroke engines of its siblings. The engine used an IHC J-4 magneto and an IHC carburetor. The little engine did not use removable sleeves, and produced about 8 horsepower on the belt and 6 horsepower at the drawbar. Fully equipped, the Cub weighed all of about 1,300 pounds.

Options

Options included a power take-off, belt pulley, adjustable front axle, electric lights and starter, swinging drawbar, and exhaust muffler.

Hydraulics

The original Cub probably didn't have the touch control hydraulic system available as an option, although it became available in 1948. An original parts book does not list hydraulics as an option. Later parts books show Touch Control hydraulics available on the first Cubs, indicating that the hydraulics were made available as a retrofit for early Cubs.

The hydraulic system on the Cub was similar to that of the Super A and Model C, although it was a one-valve system rather than the two-valve of the larger machines. These options were part of IHC's package approach to the Cub. The Cub featured a full line of implements and attachments. The unique one-armed scoop is one of the most distinctive and prized attachments, but everything from cultivators and plows to utility wagons were offered for the Cub.

The Farmall Cub remains a popular choice for collectors due to it's compact size and utility on a large yard or very small farm.

A Brief History of Styles

Researcher Jim Becker has compiled a serial number listing and brief list of style changes on the Cub. The following information was supplied courtesy of Becker. The original style Farmall Cub was built from 1947–1954. When the new Hundred Series tractors were introduced, the Cub was restyled as well and Fast Hitch became an option. A white grille was added a few years later, most likely in 1956. In 1958, the Cub was restyled again, with a new grille with large horizontal bars. That design carried through to 1963, when the grille was squared off to match another new line of International tractors. Sometime in 1964, the Farmall Cub became the International Cub and the standard color changed from red to yellow. In 1975, the Cub was redesigned, got a bit more horsepower, and was again painted red and white. That basic design was used until 1979, when the Cub was finally discontinued.

Cub Production

Model Names and Years Produced

Model	Years
Cub Lo-Boy	1955–68
Cub 154 Lo-Boy	1968–75
Cub 185 Lo-Boy	1976–77
Model 184	1978–81
Model 274 Offset	1981–84
Model 234	1982–86
Model 244	1982–86
Model 254	1982–86

Source: McCormick/IHC Archives, "Tractor Production Schedule;" Farm Tractors 1975–1995 by Larry Gay

Cub Lo-Boy

In 1954, IHC produced a lowered version of the Cub, the International Cub Lo-Boy. It was produced until 1968, with just under 24,000 built.

The eventual replacements for the Cub Lo-Boy were not offset machines, but they were a continuation of the line. The Cub 154 Lo-Boy was the replacement for the original Lo-Boy. The 154 was a standard Cub with the rear end turned 90 degrees to lower the rear end of the tractor. The front was lowered to match. Due to these changes, implements and attachments for other Cubs will not fit the Cub 154 and derivatives. The 154 was built with lawn use in mind, and the most common implements are the mowers and snow removal equipment. The Cub 154 Lo-Boy appeared in 1968, and was produced until 1974.

The replacement, the Cub 185 Lo-Boy, was produced from 1976 to 1977. A subsequent machine, the Model 184, dropped the Cub name entirely, but was built from 1978 to 1981.

Cub Production

Year	Production
1947	10,847
1948	46,483
1949	41,705
1950	21,918
1951	23,001
1952	17,829
1953	17,128
1954	7,029
1955	7,217
1956	4,573
1957	6,158
1958	7,052
1959	3,533
1960	2,408
1961	2,656
1962	1,345
1963	2,070
1964	1,657
1965	2,099
1966	2,016
1967	1,780
1968	1,976
1969	1,887
1970	1,959
1971	1,679
1972	2,075
1973	2,165
1974	2,905
1975	2,967
1976	2,214
1977	1,277
1978	1,027
1979	362
Total Prod.	252,997

Source: McCormick/IHC Archives, "Tractor Production Schedule;" Farm Tractors 1975–1995 by Larry Gay

Modern Cub Derivatives

In 1981, the line was expanded to include the 234, 244, and 254. These three tractors were in production through the mid-1980s, as Case-International continued to produce them until 1985.

The 234 was powered by a 52-cubic-inch three-cylinder diesel engine rated for 15 horsepower. The tractor came in two- and four-wheel-drive models, and the standard transmission had six forward and two reverse speeds. The 234 Hydro used a two-range hydrostatic transmission. A 540-rpm PTO, lighting, and gauges were standard, and the tractor could be equipped with a loader, snow blower, rotary tiller, harrow, several different mower decks, and front and rear blades.

The 244 and 254 used larger three-cylinder engines, 60 and 65 cubic inches, respectively, and had three-range three-speed transmissions that had nine forward and three reverse speeds. The tractors were rated for 18 and 21 horsepower, and had standard 540-rpm PTOs that could be quickly converted to run at 1,250 rpm. Both were available in two- and four-wheel-drive models. The standard three-point hitch could be equipped with hydraulic draft control.

International carried all three of these models until 1984, and Case-International sold them as the 234, 244, and 254 through 1986. In 1987, the tractors were dubbed the 235, 245, and 255, and were sold as such through 1990. In 1991, horsepower was bumped up for each model and the line was labeled the 1120, 1130, and 1140. These tractors were built through 1993.

Signs of the Future

Tractor development in the late 1950s was mainly refinement rather than the radical jumps of the 1930s, and the basic tools for the job were in place by 1954. Even in modern times, a tractor equipped with torque amplification and a live PTO can perform most of the jobs on the farm with aplomb. Due to the incredible durability and ease of rebuilds, the tractors of the 1950s could be kept at work for nearly indefinite periods of time.

Farmers could do the job well with existing equipment, so the return for purchasing a new tractor began to diminish after the mid-1950s. Selling tractors to farmers no longer brought in the tremendous profits of the 1940s and early 1950s. Tractor manufacturers had to diversify to survive, and find products that were either exploring new markets or in need of regular replacement. For IHC, the Cub was a tractor that met that need, and the Cub Cadet would become one of those products that filled both a new market and was a renewable resource that would keep selling in large numbers.

The company's forays into other markets were less successful, and cost the company more money than it made. There was, of course, the company's truck line, which was a steady producer, but the crawler and construction equipment lines were touch-and-go as far as profits. In addition, the tractor market was becoming a more difficult arena. The future for tractor companies presented some tough challenges, but IHC had an ace in the hole with the first Cub and, later, the Cub Cadet.

Cub Cadet

The Farmall Cub cracked the door to putting tractors into the hands of home owners, and the Cadet burst it wide open. The Cadet was a lawn tractor, built and designed to mow lawns and perform simple tasks around the yard. What made the Cadet special over the years was IHC's approach of bringing the features from the larger models to the little Cadet. Home owners were treated to Hydrostatic drive, power take-off, and a line of implements built especially for working the garden, plowing and blowing snow, and other home tasks.

The Cub Lo-Boy was introduced in 1954. It was available both as a Farmall and an International. Some of the Internationals were painted yellow.

The Cadet provided IHC with an almost unlimited market, and was one machine that it was able to sell in large numbers year after year. When it was introduced, the company didn't really know what it had, as IHC engineer Gordon Hershman tells it. "Ben Warren was the product manager then. He thought they oughta' do about 5,000 a year, something like that. Of course, you know, we made tens of thousands of the damn things each year. It was a bad guess, but it was in the right direction. A lotta' other stuff we had was a bad guess in the wrong direction."

The Cadet was quickly assembled, using as many common parts as possible. According to IHC engineer Harold Schram, the early versions of the Cadet didn't actually work that well. Schram helped put together the original Cadet, and stayed with the lawn and garden group for many years. "I was part of the group that was assigned the job of putting together the first Cub Cadet. The original concept was to use the fuel tank off the Cub tractor as the hood in order to commonize parts. When they put it together, it did not work that well. I think there were six or seven of us working on it. We built the first prototype about one month from the time we started."

The first version of the Cadet went into production in November 1960, but it was the second version that really made the company stand up and take notice. Schram continues, "Then the second model of the Cub Cadet came out and that turned out to be a success, more than anyone had ever dreamed. In 1962, my boss went to the Harrisburg farm show and he came back all excited about a Bolens tractor that had a U-joint

Cub Cadet Production 1961–67

Cadet (original)

1960	NA*
1961	19,091
1962	26,051
1963	14,264
Total Prod.	59,406

Cadet 70, 71, and 72

1963	2,901
1964	12,747
1965	7,434
1966	6,686
1967	3,867

Cadet 100, 102, 104, and 105

1963	1,271
1964	16,656
1965	24,779
1966	14,106
1967	12,766

Cadet 122, 123, 124, and 125

1965	43
1966	19,478
1967	27,408

Note: Although Cadet was produced until 1985, IHC records available at time of publication only list production numbers through 1967.
Source: McCormick/IHC Archives, "Tractor Production Schedule"

drive for the attachments. I was given the job of seeing what I could do with the Cub Cadet. That's when we designed a Cub Cadet where we took the belt drive out of the tractor. So it became the first garden tractor that did not use belt drive."

Gear drive helped, and Hydrostatic drive came to the Cadet in 1965. The tractor was produced in a wide variety of forms until IHC sold the Cub Cadet garden tractors to a subsidiary of MTD in 1981. Cub Cadet garden tractors continue to be built by this company, and are often available at Case-IH dealerships.

The Stay at Home Tractor

In many ways, the Cub and Cub Cadet were just as significant as the Farmall. The Farmall brought the tractor to the average farmer, while the Cub put tractors on small farms. The Cub Cadet completed the circle by bringing the tractor to the average home.

With the Cub, the small farmer found a tractor that was economical to purchase and run as well as small enough to perform all sorts of daily tasks previously reserved for a horse or man. Today, thousands of similarly sized tractors are at work on small farms, hobby farms, and estates around the country, although most of the modern machinery is foreign-made.

The Cub Cadet took the next step and brought all the conveniences of the power farm to the lawn and garden. All of a sudden, the home owner discovered a useful machine that could perform mundane tasks, fit in the garage, and didn't cost more than the family car.

Just like the Farmall, these two small tractors changed the way people viewed tractors and expanded the role of the tractor in our world. The Cub and Cub Cadet are not hailed with the same reverence as the Farmall, but they have been a part of more peoples lives than perhaps any other type of tractor.

Chapter Eleven

Polishing the Line

The Hundred Series

*"I heard stories of people who bought houses with the
money they made from overtime."*
—IH Engineer Robert Oliver, referring to the 560 recall

During the mid-1950s, the tractor market saw incremental changes that were signs of what was to come for the next 40 years.
Agricultural technology reached a plateau, with improvements
becoming more subtle. Tractors of the 1950s were efficient enough to
do most tasks on the farm, and farmers no longer needed to buy a new
tractor to farm effectively. The tractors were sturdily built, with the parts
that wore out designed to be easily replaced. With a little maintenance
and care, a Letter Series tractor could outlast the farmer.

To sell additional tractors, the industry had to meet the demand that
existed, which was for more power and accessories. This required intensive research and development, yet with decreasing sales as the reality.
The industry was putting more money into selling tractors, and selling
fewer when they were successful. By the 1960s, selling 10,000 units of a
new model in a year was remarkable.

With profit margins cut to the quick, the inevitable economic slumps proved disastrous for tractor manufacturers. Heavy equipment suffered similar woes, and sidelines such as crawlers and payloaders could not be relied on to carry companies through hard times.

The Quest for Power Begins

The demand for greater power matured in the mid-1950s. Most farmers had replaced their horses by this point. In fact, most had replaced the old machines with a more modern 1940 or later tractor. So sales of mid-sized and larger tractors began to drop off in the late 1950s.

Naturally, there was still a demand for new tractors. A certain segment of any market will always be looking the latest and greatest, and the demand was for more power and more options. The two were complimentary, as increased hydraulics required more horsepower. Forty horsepower is quite a bit when it is all devoted to driving the wheels. That becomes less than adequate when the machine has to power a hydraulic loader, turn the power take-off (PTO), and power the brakes and shifting.

The usable power of the increasingly popular diesel engines was a partial solution, and the aftermarket offered big piston and turbocharging kits. The kits brought the horsepower up plenty, but often overpowered the clutch, transmission, and so on.

The manufacturers' response was to add more options and a little more power on existing lines. By fine-tuning rather than developing new machines from the ground up, retooling and engineering costs were kept manageable, and the companies could be profitable in a tighter market. This is not to say that the new additions weren't significant. The tractor continued to become a more efficient farm tool. The roots were simply more recognizable.

Hundred Series

The Hundreds represented further refinement of the breed, with upgraded styling, a little more horsepower, and a host of new equipment. The chassis of each came from the Letter Series introduced 15 years previously, with some new sheet metal that gave the Hundred Series a cleaner look.

The Hundreds also represented complete integration of utility, wheel, and cultivating tractors. The Farmall tractors were the cultivators, while the International tractors were the utility and standard-tread tractors.

The cultivators were the standard taller-tricycle or wide-front tractors, while the utility and wheel tractors were lower to the ground with a solid, wide-front axle.

The first group of Hundreds appeared as the 100, 200, and so on, but IHC went through a phase where the name was changed every third or fourth year. The "new" machines received additional numbers (130, 240, 350, etc.), and the plethora of models made the line a bit confusing. Nevertheless, each "new" model resulted in a spike of sales, which seemed to be the driving force behind renaming the tractors each year.

The significant features of the Hundred series had already made their appearances on the later Letter Series tractors: Touch Control hydraulics, independent power take-off (IPTO), and Fast Hitch (IHC's two-point hitch). Two of these three—the hydraulics and live PTO—were improvements that would last. Fast Hitch would eventually be written out of the play by the three-point hitch, simply because it became the standard for the industry.

Live PTO simply kept the PTO turning when the clutch was disengaged. Previously, the PTO did not turn until the clutch was let out. The advantages are obvious, but the application was not that simple. Power had to be transmitted directly from the engine to the PTO, bypassing the clutch and transmission. To do so required more housings and more moving parts to wear out, but the results were well worth the effort.

Fast Hitch Development

Fast Hitch was IHC's attempt to answer the Ferguson three-point system on the Fords. Ferguson's system was effective and popular. After Ford and Ferguson split up, the three-point became available to any manufacturer, but it wasn't available for free. The long tradition of IHC using in-house equipment—carburetors, magnetos, and so on—was not about to be dropped to use a feature that was developed on the Ford.

Fast Hitch was the company's response. The engineers developed it to solve the Ferguson system's problem of erratic draft. The plow wouldn't come right out of the ground, but it would vary in depth as soil conditions changed. Fast Hitch solved that problem, but the buyer didn't seem to care.

In 1945, Gordon L. Hershman came to IHC from the military, where he worked on hydraulics for firing rockets from ships in World War II. He ended up working on Touch Control hydraulics and then developed a hitch to compete with the Ferguson system. He explained that the engi-

Farmall and International 100, 130, and 140 Production

Farmall 100

Year	Production
1954	?*
1955	9,990
1956	8,585
Total Prod.	18,575

Serial numbers are listed for 1954, but production schedule shows none built that year. Models were built, but the production sheet simply records it a bit differently.

Farmall and International 130

Year	Production
1956	?*
1957	6,946
1958	2,821
Total Prod.	9,767

Serial numbers are listed for 1956, but production schedule shows none built that year. Models were built, but the production sheet simply records it a bit differently.

Farmall and International 100, 130, and 140 Production (cont.)

Farmall and International 140†

Year	Production
1958	1,386
1959	5,183
1960	3,678
1961	5,188
1962	4,480
1963	3,844
1964	3,157
1965	2,810
1966	3,532
1967	2,620
1968	2,554*
1969	2,394*
1970	2,124*
1971	2,181*
1972	1,902*
1973	2,213*
1974	4,003*
1975	3,050*
1976	3,066*
1977	2,272*
1978	1,433*
1979	NA
Total Prod.	63,070*

Production numbers from 1968-on were calculated by subtracting serial numbers, which is a good estimate but not exact.
†These numbers come up about 973 short of other totals. The 1958 to 1967 production of the International 140 may not be included in IHC documentation.
Source: 1958–67 from McCormick/IHC Archives, "Tractor Production Schedule"

neering group felt that the Ford-Ferguson hitch provided erratic draft control, which meant that fields were worked unevenly and inefficiently. "If you were plowing a field and got into heavy going, it came out of the ground. If you were trying to plow 8 inches deep, and you were in different soils in a field, the damn thing would be 8 inches sometimes, and other times it'd be 4 inches. We didn't think that was much good. That's why we came up with the Fast Hitch."

Besides the advantages of improved draft control, Fast Hitch was also much easier to hook up. Engineer Robert Oliver also worked on developing hitches, and recalls that Fast Hitch was well received because of easy access. "The tools were getting bigger and bigger, and it was much easier to hitch up than to pound those links on with your heel, or reef those

The Hundred Series tractors reflected the first major revision of the International trac- tor line since 1939. The Hundred Series models were introduced beginning late in 1954.

tools around to get 'em lined up. With Fast Hitch, you could just back in, pick up, and go. It was very popular."

Despite the advantages of Fast Hitch, the buyer still preferred the three-point hitch, so IHC looked to develop one. The company looked closely at existing three-point hitch designs, and built its own system with an eye toward avoiding the weaknesses of the Ferguson hitch.

The improved hitch featured Traction Control, which allowed the operator to vary the amount of ground pressure on the rear tires with four different settings. The system was mechanical, and automatically matched ground pressure to load in any of the four settings. The prob- lem of erratic draft control was solved. "It could go over the terrace and the tool never left the ground. It had the best regulation curve in the industry," Oliver said.

Fast Hitch was slowly phased out as the three-point hitch became an industry standard. International had a three-point of its own, and it was one of the better systems available. Business as usual.

100 Series

The 100 Series were basically Farmall Super As with a few new features, slightly more power, and new badges. The tractor was built as the 100 from 1954 to 1956, as the 130 from 1956 to 1958, and as the 140 from 1958 to 1979. Although the machine was refined and restyled along the way, the chassis and engine design was not significantly changed in an amazing production span of 40 years.

The Farmall and International 100 engine was the venerable C-123 unit used in the Super C and Super A-1. It was equipped to burn gasoline, which was the standard by that time, but could be equipped to burn kerosene or distillate. The 100's engine produced 18 belt and 14 drawbar horsepower. The Farmall 130 engine put out a little more power, at 20 belt and 16 drawbar horsepower. Although horsepower kept climbing incrementally over the years, the same bore and stroke was used.

Power was transmitted with a 9-inch dry-disk clutch, and the transmission was a four-speed. The same replaceable cylinder sleeves and ball and roller bearings were used, as well. The rear wheel tread was adjustable from 40 to 68 inches, and the optional front was adjustable from 44 to 64 inches. The tractor weighed about 2,600 pounds.

By 1958, the cooling system used pump circulation and was thermostatically controlled, leaving the Cub as the last in the line to use a thermosyphon system. Earlier tractors (A, B, Super A, Super A-1) used thermosyphon cooling with an optional booster pump.

Options and Attachments

The 100 Series came from the factory with a bevy of standard equipment that included Fast Hitch, Touch Control hydraulics, fenders, nonadjustable front end, battery ignition, lights, and a muffler.

Other options included a belt pulley, PTO, rear light and tail light, exhaust valve rotators, foot accelerator, radiator shutters, wheel weights, and a host of custom implements and drawbar attachments. A Farmall 100 Hi-Clear was available, as well, with the same host of options and attachments.

Hydra-Creeper

One of the more interesting options was "Hydra-Creeper," which allowed the tractor to crawl along at speeds of 1/4 to 1 mile per hour at full throttle. This attachment was used for transplanting crops or for pulling equipment when harvesting, all primarily for smaller vegetable, flower, or tobacco farms. The unit consisted of a hydraulic pump that was driven by the PTO. The hydraulics drove a chain and sprocket that were geared extremely low. With Hydra-Creeper engaged, the regular transmission was used, but speeds in each gear ranged from about 0.2 miles per hour in first to 1.1 miles per hour in fourth. Hydra-Creeper could be disengaged at will, allowing normal operation of the transmission. To use the PTO, the unit had to be disconnected.

The 100 Series never lit the world on fire with sales like the Model A and B, but they sold steadily over an unrivaled span of years, with nearly 100,000 machines rolling out the door. These tractors are an ideal size for a hobby farm or residence, and have become increasingly popular with collectors as well.

200 Series

The 200 Series was the upgraded version of the Super C. The Super C already had Touch Control hydraulics, so the 200 was not really much of a change. The 200 was produced from 1954 to 1956, the 230 was produced from 1956 to 1958, and the 240 was produced from 1958 to 1962. Both Farmall and International 200 Series tractors were produced, the distinction being the Farmall was the cultivating tractor and the International was the utility tractor.

The 200 series used the same C-123 engine as the 100 Series, but the 200 put out 22 belt and 16 drawbar horsepower. Where the

Farmall and International 200, 230, 240 Production

Farmall and International 200

Year	Production
1954	?*
1955	7,576
1956	6,150
Total Prod.	13,726

Serial numbers are listed for 1954, but production schedule shows none built that year. Models were built, but the production sheet simply records it a bit differently.

Farmall and International 230

1956	?*
1957	5,480
1958	1,973
Total Prod.	7,453

Serial numbers are listed for 1956, but production schedule shows none built that year. Models were built, but the production sheet simply records it a bit differently.

Farmall 240

1958	1,232
1959	1,481
1960	861
1961	136
Total Prod.	3,710

International 240

1958	3960
1959	3,949
1960	1,663
1961	671
1962	86
Total Prod.	10,329

Source: McCormick/IHC Archives, "Tractor Production Schedule"

Farmall and International 300, 350, 330, and 340 Production

Farmall 300

Year	Production
1954	?*
1955	19,466
1956	10,528
Total Prod.	29,994

A limited number of serial numbers are listed for 1954, but production schedule shows none built that year. Models were built, but the production sheet simply records it a bit differently.

International 300

1955	14,312
1956	18,864
Total Prod.	33,176

Farmall 350

1956	?*
1957	11,855
1958	4,950
Total Prod.	16,805

About 500 serial numbers are listed for 1954, but production schedule shows none built that year. Models were built, but the production sheet simply records it a bit differently.

International 350

1956	?*
1957	13,575
1958	4,425
Total Prod.	18,000

About 300 serial numbers are listed for 1956, but production schedule shows none built that year. Models were built, but the production sheet simply records it a bit differently.

International 330

1957	?*
1958	4,261

About 950 serial numbers are listed for 1957, but production schedule shows none built that year. Models were built, but the production sheet simply records it a bit differently.

Farmall and International 300, 350, 330, and 340 Production (cont.)

Farmall 340

1958	2,000
1959	2,574
1960	1,433
1961	1,221
1962	182
1963	77
Total Prod.	7,487

International 340

1958	1,927
1959	2,852
1960	2,889
1961	2,963
1962	988
1963	435
Total Prod.	12,054

Source: McCormick/IHC Archives, "Tractor Production Schedule"

increased horsepower came from in the C-123 engine is not clear, although piston speed is listed in an IHC brochure as being higher, making it likely that additional rpm yielded the increased power of the 200. The engine cooling was thermostatically controlled, and it was equipped to use gasoline with kerosene or distillate equipment optional. Horsepower increased slightly as the model progressed through the years.

The 200s were equipped with IHC's new double-disc brakes, which were enclosed and mechanically operated. Other standard equipment included a temperature gauge, lights, and electric starting, Touch Control hydraulics, Fast Hitch, muffler, and rear tread width adjustable from 48 to 88 inches. Options include the Hydra-Creeper, combination PTO and belt pulley, hydraulic remote control, hour meter, and a high-ratio third gear.

Touted as ideal for a farm of 80 to 120 acres, the 200 filled the niche of an affordable smaller tractor with enough power to pull two 14-inch plows and perform all the tasks necessary on a smaller farm. Lifetime sales of the 200 Series were over 30,000 units, and ended in 1961 for the Farmall 240 and in 1962 for the International 240.

300 Series

The 300 Series was the descendant of the Model H and the Four Series. Like the rest of the Hundreds, new sheet metal, more power, and more options were the order of the day. These mid-sized tractors were available with a Torque Amplifier (TA) and "live" or independent PTO.

The 300s were built from 1954 to 1956, with the 350 built from 1956 to 1958. The 340 was built from 1958 to 1963. An International 330 (but not a Farmall 330) was built in 1957 and 1958. As with the smaller tractors, the Farmalls were the cultivating tractors and the Internationals were the standard-tread utility tractors.

Horsepower was up from the Super H, with 34 belt and 27 drawbar horsepower when burning gas. The engine could be equipped to burn kerosene, distillate, or LPG. The 300's C-169 four-cylinder engine had a slightly larger bore than the Super H's C-164 engine.

Continental Engine in Model 350

According to an article in *Antique Power* magazine, a Continental diesel engine was available for the 350. IHC documentation shows a 193-cubic-inch diesel for the 350, and a 166-cubic-inch diesel engine for the 340, indicating a probable switch to an International engine for the 340. The 166-cubic-inch engine was also used in the TD-340 crawler, increasing the odds of the engine being an IHC-built unit.

Cooling was the standard thermostatically controlled water pump system, and the engine was lubricated by an oil pump. The tractor weighed in between 4,000 and 5,000 pounds, depending on equipment.

Torque Amplifier

The transmission on the 300 could be ordered as the standard five-speed, or with Torque Amplification (TA), which provided ten effective forward speeds. A gear reduction system was used, and the farmer could shift on the fly between two slightly different gear ratios, giving two ranges to each gear. This worked great for plowing in soil or varied consistency, or for any other situation that required a quick, subtle change in speed.

The new TA was a part of a new concept to tractor technology; matching load to speed. Research had shown that tractors work most effectively when power exerted is matched exactly to what is required. Anyone who has plowed with an older tractor where third gear is too high and second is too low understands this need. More gear choices make it more likely that load can be perfectly matched and the result is less time in the field and better fuel efficiency.

Standard equipment was similar to the previous 100s and 200s, with electric starting and lights and so on installed. The independent power take-off (IPTO) was another important feature for the 300s.

The 300s sold well for IHC, selling about 35,000 in 1955 alone with more than 100,000 sold from the introduction in 1954 to the close of production in 1963. Like the Farmall H that preceded it, the 300 Series were venerable, useful machines that were used on a wide range of farms and applications.

400 Series

The successor to the Farmall M and Six Series was the 400 Series. These tractors used the same engine as the Super M and its siblings, and featured the IPTO and TA that debuted on the Super M. The 400s had the new sheet metal and more power as well. They also featured the option of burning liquefied petroleum gas (LPG).

The 400 Series four-cylinder engine was the C-264 from the Super M, although it had more snort, with the standard gasoline burner putting out 45 belt and 36 drawbar horsepower. The engine could be equipped to burn kerosene, distillate, or LPG, and diesel version was optional.

Model 600 and 650

Produced for only a few years in the late 1950s, the big 600 and its successor, the 650, were IHC's largest, most powerful wheel tractors. They were also simply a rebadged Super Nine with a bit of new equipment. According to IHC records, the 600 and 650 were produced only as standard tread Internationals, with no Farmall models. The 600 and 650 used the big 350-cubic-inch engine from the Super Nines, and were rated for 60 belt horsepower. Weighing in at over four tons, the 600 also boasted all of IHC's latest innovations—IPTO, power steering, and an optional cigar lighter.

The 350-cubic-inch four-cylinder engine had coffee can-sized pistons, with a 4 1/2-inch bore and 5 1/2-inch stroke. The standard engine burned gasoline, but it could be equipped to burn kerosene or distillate. Diesel and LPG versions were available as well. The pistons were aluminum, and the cylinder sleeves were replaceable. The diesel engine used fuel injection and had a compression ratio of 15.6:1. Battery ignition was standard, as were lights and an electric starter.

An International 650 was tested at Nebraska in 1957, and it recorded 55 belt and 43 drawbar horsepower. The horsepower figures are respectable, but the number that best describes the power of the 600 and 650 is the 530 foot-pounds of torque produced by the powerful, low-revving engine. That's as much torque as a big-block V-8 engine.

Farmall and International 400 and 450 Production

Farmall 400

Year	Production
1954	?*
1955	24,440
1956	16517

About 2,000 serial numbers are listed for 1954, but production schedule shows none built that year. Models were built, but the production sheet simply records it a bit differently.

International W-400

1955	1,336
1956	1,932
Total Prod.	3,268

Farmall 450

1956	?*
1957	17,852
1958	7,737
Total Prod.	25,589

About 1,200 serial numbers are listed for 1956, but production schedule shows none built that year. Models were built, but the production sheet simply records it a bit differently.

International W-450

1956	?*
1957	1,091
1958	718
Total Prod.	1,809

About 60 serial numbers are listed for 1956, but production schedule shows none built that year. Models were built, but the production sheet simply records it a bit differently.
Source: McCormick/IHC Archives, "Tractor Production Schedule"

The test tractor burned gasoline, and weighed more than 10,000 pounds with wheel weights. In the low-gear maximum pull test, the 650 snorted over 8,500 pounds of force—more than 80 percent of its weight.

The engine was cooled with a thermostatically-controlled radiator and fan system. The five-speed transmission was mated to a 12-inch single-plate clutch. Top gear was for road use, and ran the 600 up to over 15 miles per hour. Disc brakes and 14 x 34-inch rear tires were standard.

Optional equipment included the IPTO, Hydra-Touch hydraulic system, belt pulley, foot-operated throttle, rice field tires, magneto ignition, high-altitude pistons, radiator shutters, tachometer, hour meter, wheel weights, heavy duty rear axles, and a cigar lighter (real farmers who used tractors with over 530 foot-pounds of torque apparently smoked cigars rather than cigarettes).

International 600 and 650 Production	
International 600	
1956	1,484
1957	32
Total Prod.	1,516
International 650	
1957	2,561
1958	2,372
Total Prod.	4,933

Source: McCormick/IHC Archives, "Tractor Production Schedule"

Production Figures

Several IHC production sheets clearly indicate that about 2,000 machines were sold each year (see chart for specifics). Note that the 600 and 650 were both produced in 1957, although only a few 600s were built that year.

Considering the fact that the 600 and 650 were essentially Super Nines, the figure becomes even more plausible. The Super Nine sold a few thousand per year consistently, as did the International 660 that replaced the 650. The serial number references are quite confusing, but the production figures in IHC records are quite clear.

With less than 7,000 units produced, the 600 and 650 are among the rare birds of the line. The lack of information and the odd serial numbers—especially on the 600—lend a veil of mystery to these powerful tractors.

New Forty and Sixty Series

The next step for International was the first major departure from the basic styling of the Letter Series since 1939, nearly 20 years prior. It was 1958, and the new IHC tractors featured a square front end with a stacked-bar grill. The look was strikingly different. It would also be the styling that defined IHC tractors almost until the end. The tractors' appearance would become sleeker as time passed, but the tractor of 1980s didn't really look that much different from the tractors of 1958. Option choices were up, especially comfort options, and hydraulic systems were increasing in capacity.

The Farmall 340 was built from 1958 to 1963, and featured an all-new look for IHC tractors as well as Fast Hitch. The hitch system was the IHC answer to the Ferguson three-point system. The three-point became standard and Fast Hitch faded away.

Forty Series

The Forty Series tractors were the latest editions of the Hundred Series introduced in the mid-1950s, although they sported the new look and some new bells and whistles, including the new three-point Fast Hitch. A 140, 240, and 340 were offered as Farmall cultivating and International standard tread tractors.

Sixty Series

The Sixty Series tractors—the 460, 560, and enigmatic 660—were the real news in the new lineup. While the 460 and 560 were introduced in 1958 and the 660 in 1959, they shared a common design. All three used new IHC six-cylinder power plants, and bore the new square styling. More importantly, all three put up respectable numbers of horsepower. The horsepower race was heating up, and the Sixty Series was IHC's entry.

The standard engines burned gasoline, and the usual kerosene and distillate equipment was optional, with diesel and LPG versions available as well. The engine boasted a host of new features, as well. A wedge-shaped combustion chamber and angled valves improved flow for more horsepower and better fuel economy. A 12-volt electrical system was used, and diesels had a new injection pump and glow plugs in each pre-combustion chamber. All of the machines had the Torque Amplifier coupled to a five-speed transmission, yielding 10 forward and 2 reverse speeds. The tractors were equipped with an independent 540-rpm power take-off.

New Hydraulics

The hydraulic system was upgraded, with the option of a one-, two-, or three-valve Hydra-Touch. The hydraulic system was housed in the rear end, with several different pumps available. The hydraulic system was fitted with a replaceable filter to keep the fluid clean, and the hydraulics were "live," meaning power was supplied whenever the engine was running.

The 460 and 560 were available as Farmall cultivating and International standard tread machines, while the big 660 came only as an International standard tread.

Farmall and International 460

The 460 used a 221-cubic-inch six-cylinder gasoline engine that produced 44 belt and 36 drawbar horsepower at 1,800 rpm. The 460's six-cylinder diesel engine displaced 236 cubic inches and produced the same amount of horsepower in both tests, also at 1,800 rpm. The tractor weighed about 6,000 pounds and sported a five-speed transmission that could be equipped with Torque Amplifier (TA).

Farmall and International 560

The standard gasoline engine in the 560 displaced 263 cubic inches and was good for 61 belt and 45 drawbar horsepower at 1,800 rpm. The same engine equipped for propane made about the same horsepower on the drawbar, but slipped to 53 belt horsepower. The 282-cubic-inch six-cylinder diesel put out 59 belt and 44 drawbar horsepower at 1,800 rpm. All versions weighed over 6,500 pounds.

Problems with the 560

Unfortunately, the high horsepower of the 560 overpowered its transmission. The problem was fixed, but the company had to scramble to recall the machines out in the field and to fix the unsold machines that were off the production line.

To fix the problems with the 560, International applied a warranty program of a scope unrivaled in company history. Engineer Robert Oliver recalls the stories about the 560 warranty effort. "They had a crew of people work around the clock to repair tractors. They had the complete field at Hinsdale filled with those tractors off the production line. They worked around the clock, unlimited overtime. I heard stories of people who bought houses with the money they made from overtime." The problem was cured with later models and earlier ones were retrofitted, but the damage was done to IHC's nearly spotless reputation.

The 660

The big 660 appeared in 1959. It was the replacement for the Model 650, and used the new six-cylinder engine and Planetary Power hubs in an evolution of the W-9 chassis. It weighed nearly 10,000 pounds, and soared to over 15,000 pounds with front and rear ballast. The gasoline engine was the same 263-cubic-inch unit from the 560, but it ran at 2,400 rpm. The result was 81 PTO and 70 drawbar horsepower when tested at Nebraska in 1959.

The diesel engine was the 281-cubic-inch model with an

Sixty Series Production	
Farmall 460	
1958	3,851
1959	10,574
1960	7,444
1961	5,097
1962	3,637
1963	2,048
Total Prod.	32,651
International 460	
1958	2,072
1959	3,823
1960	2,988
1961	2,152
1962	537
1963	19
Total Prod.	11,591
Farmall 560	
1958	6,161
1959	18,063
1960	10,622
1961	11,296
1962	11,650
1963	7,794
Total Prod.	65,586
International 560	
1958	671
1959	1,653
1960	1,077
1961	1,034
1962	585
1963	613
Total Prod.	5,633
International 660	
1959	2,666
1960	1,100
1961	1,529
1962	1,030
1963	676
Total Prod.	7,001

Source: McCormick/IHC Archives, "Tractor Production Schedule"

The new Farmall 460 and 560 were the high-profile machines in the new line for 1958. Both were designed with the horsepower-hungry times in mind, and used new six-cylinder engines.

18:1 compression ratio. Power output was similar to the gasoline version, with 79 PTO and 65 drawbar horsepower. An LPG engine was also available, and it posted numbers identical to the gasoline version, at 81 PTO and 70 drawbar horsepower.

Sadly, the Sixty Series' biggest impact was the problems with the 560. Like the transmission problems suffered on the TD-24, the problems with the 560 sullied the reputation of an otherwise great tractor. In both cases, International had hurried the machine to the showroom, and suffered because of it. As the 1960s dawned, International's 30-year juggernaut was on rocky ground. It remained one of the powers of the industry, but the errors had tainted the integrity of the IHC label a bit.

Chapter Twelve

Powering Into Modern Times

Surviving the Second Age of Consolidation

"Like the Church, the Monarchy and the Communist Party in other times and places, the corporation is today's dominant institution."
—The Corporation, *2003 documentary film*

The 1960s were pivotal for the tractor industry. John Deere, Case, Oliver, and IHC all introduced new tractor lines early in the decade, and the competition was fierce. John Deere and IHC were the big players, but the market continued to shrink from the late 1940s and early 1950s, when manufacturers could sell tens of thousands of a full-sized model year after year.

For full-size models, the news for the 1960s was again power and add-ons. Hydraulics began to play a huge role in tractor technology, and International's new 706 and 806 tractors boasted no fewer than three hydraulic pumps. Everything from steering to brakes became hydraulic, and horsepower requirements continued to spiral.

The increased demand for power made turbocharging a common practice for tractor manufacturers, and four-wheel-drive became a more significant feature. It was an exciting time for IHC, with the first new-from-the-ground-up tractors appearing since the introduction of the Letter Series in 1939. The industry in general was on the move, and International was poised to remain one of the leaders.

Beginning in about 1970, the tractor makers went through a raft of consolidations, sales, and failures that continues today. Where once there were hundreds of tractor manufacturers, only a handful remain. Through the 1970s and early 1980s, IHC struggled through a series of economic fluctuations. In 1985, the struggle ended when the agricultural division was sold to Tenneco and merged with the Case tractor division. The International name survived on the "Case-International" badges on the "new" tractor line, but it wasn't even close to the same.

404 and 504

The first new weapons for IHC were the 404 and 504, two mid-sized machines introduced in 1961 that had a number of improvements. The tractors were offered with an IHC-engineered draft control. The system was unique to IHC, and worked well, but would be outsold and pushed off the market by the Ferguson System. Other innovations of the 404 and 504 included an oil cooler and a dry-type air cleaner. The oil cooler was necessitated by the increasing horsepower of modern tractors, and became standard equipment for most tractors. The dry-type air cleaner simply made sense, just as the paper filters in automobiles do today. Another advance on these tractors was what IHC called Hydrostatic power steering. By using hydraulics to turn the front wheels, awkward, bulky mechanical linkages were eliminated. Both models were available in gasoline or LPG setups, and the 504 was available with a diesel engine as well.

Both of these sold modestly, with total sales of more than 10,000 when they were discontinued late in the 1960s. For this model, the standard tread configuration was more popular than the row crop version.

Getting More Power To the Ground

The largest demand on the farm was still more power and, in the 1960s, two developing technologies brought that about: turbocharging and four-wheel drive. The two fed off each other, really. As the demand for more options and more power rose, the need for greater traction tagged along. It doesn't take a rocket scientist to see that turbocharging brought more power and four-wheel-drive put it to the ground more effectively.

Turbocharging is an effective way to increase power in tractor engines. Tractor engines run at low rpm with heavy loads, so traditional hot rodder's methods of increasing high-rpm horsepower are not effective on tractor engines. More efficient flow is a viable option, but increased complexity is undesirable for a machine that needs to run hour after hour without servicing.

Turbocharging pressurizes the intake with an impeller driven by exhaust gas. The drawback to turbocharging is poor throttle response. In a car or motorcycle, you can feel a bit of lag as the impeller gets up to speed when you punch the accelerator. A tractor engine runs at constant speeds, so throttle response is really not an issue. Turbochargers provide engines with dramatic boosts in horsepower, and allow the machines to maintain reliability and decent fuel economy.

The rising power made traction more and more of an issue. Farmers were pulling larger implements, and doing more in less time. Dual rear tires helped, but they were somewhat unwieldy and clumsy to mount. Besides, the nearly 100-horsepower of the machines was at times more power than two driven wheels could get to the ground. Four-wheel-drive was a logical solution, but it presented a tricky engineering task.

One solution came from Steiger, a company that formed in 1957 with the express purpose of building a high-horsepower tractor. Steiger found success building powerful, articulated, four-wheel-drive tractors, and some manufacturers bought machines from Steiger and rebadged them as their own. The situation was less than ideal from a profit standpoint, but allowed companies to put a big four-wheel-drive on the lot without much research and development.

4300

The big 4300 was just such a solution for IHC. The machine was an IHC product, but it was built for the construction division and was too big for most fields. It was brought over and sold as an agricultural tractor beginning in 1961. The model provided IH with bragging rights, as it was one of the most powerful machines on the market.

The mighty 300-horsepower six-cylinder engine was an 817-cubic-inch turbocharged IHC unit. The result was a monstrous tractor that was also the first IHC tractor to breathe through a turbo. The hydraulic drive used a torque converter and six-range powershift transmission. A 10-bottom plow was developed especially for the 4300, and with a total ballasted weight of more than 30,000 pounds, the 4300 was plenty capable of pulling the big plow.

The 4300 helped fill a hole in the IHC line, for the time being. The company had a couple of new models up its sleeve before finding a better solution for the big four-wheel-drive problem.

706 and 806

The 706 and 806 were huge successes for IHC. The two new models were designed from the engine housing back to handle high horsepower output. The tractors were introduced in 1963 and used hydraulics to power many of the systems, from brakes and steering to auxiliary equipment like front-end loaders. The hydraulics required large amounts of horsepower, and the 706 and 806 put out nearly double the amount of power produced by the big tractors of the late 1940s. These two backed up the big numbers with all the latest options, from power steering and brakes to 540- and 1,000-rpm IPTOs and a full bevy of gauges.

International brought out a number of significant tractors in the 1960s, but the 706 and 806 were the most successful. The high-horsepower tractors breathed new life into International and were billed as high-speed tractors.

Farmall and International Models

As with previous IHC tractor lines, both International standard tread and Farmall row-crop versions were available. The trend at the time was moving away from the tricycle-style cultivating tractors, but the sales of the 706 and 806 lines did not reflect that. This may have been because the standard tread Farmall included two auxiliary hydraulic valves and a three-point hitch, while the row crop International had only one auxiliary valve and a swinging drawbar as standard equipment. The Farmalls were also longer and wider than the International models, with a greater range of rear tread width adjustment.

706 Engine

The 706 could be equipped with the D-232 diesel or the C-263 gasoline engine. The gas engine could be outfitted to burn LPG, and all were six-cylinders. Both the gas and diesel model were rated for 74 PTO and 65 drawbar horsepower.

806 Engine

The 806 could also be equipped with a gas, diesel, or LPG six-cylinder engine. The gas or LPG unit used the big C-301, and the diesel was the D-361. Horsepower for both was rated at 95 PTO and 84 drawbar.

New Transmission

The transmission for the 706 and 806 was a dual-range four-speed, giving eight forward speeds without the optional Torque Amplifier (TA). The addition of more gear selections and a wider effective engine rpm range (from 900 to 2,400 rpm on the 706 and 806), allowed the tractors to match power output to load as never before. With the optional TA, the 706/806 tractors had 16 effective forward speeds.

Evolution of the PTO

The independent power take-off (IPTO) was more or less standard by the mid-1960s. The latest development was adding the 1,000-rpm PTO to the standard 540-rpm PTO. The new tractors featured both speeds on the tractor, as the 1,000-rpm PTO was the newest addition, but many implements still required the 540-rpm PTO. One of the problems with the early IPTO was that it engaged too suddenly. This was solved on the 706 and 806. When the PTO was engaged with the clutch, it was naturally engaged progressively as the clutch was let out. The new IPTO was

The International 4100 was a quick solution to the problem of putting a four-wheel-drive farm tractor into the line. The 4100 was designed, tested, and put into production in an amazing 23 months.

engaged by a lever, and spun suddenly from zero to the rated rpm. The result was bent and broken implements. The solution IHC promoted on the 706 and 806 was to hydraulically activate the PTO, which gave a gradually engagement.

Hydraulics

Hydraulic power was used to power the brakes, steering and other systems, and the tractor used three hydraulic pumps. A 12-gallon-per-minute (gpm) pump in the main frame operated the hitch and mounted equipment. A 9-gpm pump in the clutch housing operated the power steering, power brakes, and TA system. A 3-gpm pump powered shifting and lubricated the IPTO. Draft control was hydraulic as well.

Comfort and Convenience

The 706 and 806 were engineered to appease the operator, with more comfortable seats, shielded operator platforms, body panels that flipped up for easy servicing, and a full complement of gauges. Both could be ordered with factory front-wheel assist, making all four wheels driven.

The mid-1960s were an exciting time for IHC. The early embarrassment of the flawed 560 was put behind, and the new 706 and 806 erased any doubts about the company's ability to produce cutting-edge agricultural equipment. Although these tractors met the demand for more features and more horsepower, there was another challenge waiting for the company, one that it wouldn't solve to the staff's satisfaction until the introduction of the stunning 2+2 models in 1979.

4100

The 4300 was IHC's only big four-wheel-drive, and it wasn't filling the bill. Converted from the construction equipment line, it was not designed for the needs of the farmer. In 1963, with the new 706 and 806 at the dealerships, the company looked at building a big four-wheel-drive tractor designed specifically for farm use.

Rapid Development

With demand up for a big four-wheel-drive capable of working the fields, International pushed the 4100 to market as rapidly as possible. The resulting development work was amazingly speedy, with design work beginning in August of 1963 and the first production tractor off the line in July of 1965.

Harold Schram, an IHC engineer, worked on the design team for the 4100. "I was asked to layout a four-wheel-drive tractor with approximately 150 horsepower. I started with the engine—of course, we always used IH engines. The ET429 was about that horsepower, so I started with that.

"We could not afford to design a new transmission . . . we finally decided that we could stretch the transmission that was used on the 706 and 806. We did it by putting a reduction after it so there was less torque going into the transmission.

"The big thing at that time was independent drive axles that were used on wheel loaders. We picked up a couple of those. We looked at Clark and Rockwell Standard. We chose a steerable wheel system. We had a choice, because Steiger and Versatile had an articulated system and Case had a steerable system. We went with the steerable system, because we thought it was more compatible for use in row-crop situations.

"Basically, once you make those decisions, then you've got to have enough room to get an entrance [to the cab] up so you put your engine and transmission components in first and find out where you can set your axles and then you go up, building an operator's deck, fenders, making sure you've got enough space for both sets of wheels to turn and it just sorta' builds up. We went from there to building experimentals."

The time period from the drawing board to production was incredibly short and most of the hard work was done by only 10 people. The original Farmall took more than five years of refinement to build, and modern tractors require a similar amount of time and testing. Schram went on, "We put that tractor into production in under twenty-three months, from the ground up. It was a brand new tractor in a brand new market that Harvester had never been in before. We started August 15th of 1963 and the first production models rolled off in July of 1965."

Sorting Out the Bugs

There were a few problems with the 4100, perhaps a natural result of the accelerated development. Schram said, "There were some problems, basically in the steering system. The major ongoing problem was with the rear-wheel steering. We had a linkage system in there and did not have a check valve in it originally and the wheels would tend to wander a little bit at road speed. Then we put a check valve into it, so we weren't just depending on the linkage. That held the rear wheels steady."

Engineer Robert Oliver was working on the test group when Schram was working on design. He remembers testing prototype 4100s, which had a tendency to break the clutch input shaft.

"We had some real experiences in that," he said. "We could break [the clutch input shaft] in like 10 feet. It took you all day to get a new one put in. [The problem] ended up being a resonance problem. We had to change the spring constant in the shaft to keep the resonance out of it. Once we got that out of it, it didn't break anymore," Oliver said.

Engine and Late Models

The 4100's engine was a 429-cubic-inch, six-cylinder turbocharged diesel, and the cab boasted one of the first factory installations of a heater and air conditioner. The original 4100 was yellow and white, and the paint scheme went to red and white in 1972 with the 4166. The 4100 and various versions of it didn't set the world on fire with sales, but they remained in the line through 1978. Its replacements, the 3388 and 3588 "2+2" machines, would be the revolutionary four-wheel-drive tractor that IHC was searching for all along.

Other 1965 Models

Other new-for-1965 models included the 1206 and the 656. The 1206 was a larger two-wheel-drive tractor, the first Farmall to be rated for over 100 PTO horsepower. The engine was a turbocharged, 361-cubic-inch six-cylinder diesel. The tractor had a 16-speed transmission. It was produced from 1965 to 1967, and was rebadged as the 1256.

These tractors could be equipped with a factory cab with heat and air conditioning. This was the first appearance of a cab on two-wheel-drive International tractors. The cab was built by Stolper-Allen of Menomonee Falls, Wisconsin.

The 656 was also introduced in 1965, and production ran until 1973. The tractor was powered by a six-cylinder turbocharged 281-cubic-inch diesel that produced 64 PTO horsepower. It was fit in the line just below the 706, and used a similar design. In 1967, the 656 was available with International's new hydrostatic transmission. The HT-340 Gas Turbine Tractor had been one of the first experimentals to carry the hydrostatic transmission, and the 656 was the first production tractor to use it. The hydrostatic transmission used hydraulics to transfer power from the engine's flywheel to a standard final drive. Speed was controlled with an infinitely-variable single control known as the speed ratio (S/R) lever. The operator could choose from high and low range, and set the speed in each range with the S/R lever.

The advantage to hydrostatic was that speed could be fine-tuned to a greater degree than with standard transmissions. The disadvantage was that the hydraulic drive transferred power less efficiently and required more engine horsepower for equal drawbar pull.

The 656 was joined by several updated models that featured some new features and paint schemes. The 706 and 806 became the 756 and 856, while the 1206 was renamed the 1256. The optional cab for these tractors was changed to a model built by Excel Industries that had more operator room.

The Seven-Year Itch

By the 1970s, it took a peak year of tractor sales by the entire industry to top—just by a hair—IHC's total tractor production of over 170,000 tractors in 1951. The industry had changed, and tractor manufacturers simply could not sell the numbers possible in the past. The number of farms and farmers had declined, and the advantages of adding a new tractor to the farm had diminished.

Tractor sales seemed to cycle and peak about every seven years. With a peak in 1966, the next was appeared on schedule in 1973. A total of 196,994 tractors were sold that year, the peak for the decade as it would turn out. Sales would peak again in 1979 at 188,267 units. As a side note, the Farmall name was dropped from the International line in 1973, suspending a legacy that began nearly five decades earlier.

The tractors of the 1970s were going through the same changes as those of the 1960s, only on a larger scale. Horsepower was again climbing, and manufacturers turned to big-cubic-inch turbocharged and intercooled diesel engines for horsepower that went as high as 350. Front-wheel assist became a common option on most mid-sized to large tractors, and four-wheel-drive was the rule for the big machines over 150 horsepower or so. Even smaller models began using four-wheel-drive, and the small four-wheel-drive tractor would become one of the growing market segments of the decade.

Foreign makers became increasingly competitive in the 1970s, especially with smaller tractors. In fact, it became more common for small tractors to be built overseas than domestically.

1969

In 1969, International had one new model and some new names for familiar faces, most of which were available with hydrostatic transmissions. The 856 became the 826, and was available in both gear-drive and hydrostatic transmissions. The hydrostatic version was rated for about 20 horsepower less than the gear-drive machine. Gasoline, LPG, or diesel models were available. The 1056 became the 1026, and was also available with a hydrostatic transmission. The 4100 became the 4156, a tractor that used the chassis design from the 1056 and a larger, more powerful six-cylinder diesel engine. The turbocharged 407-cubic-inch powerplant put out 131 PTO horsepower through the gear-drive transmission. It was built until 1971, when it was rebadged as the 1466.

66 Series (1971)

International's next big push for new models came in 1971, when the company introduced 11 new models. Most of the new tractors were in the 66 Series, with seven tractors ranging from 80 to 133 PTO horsepower. All used six-cylinder engines, and most were available with either hydrostatic or gear-drive transmissions. The biggest of the new models was the 1468, which was powered by a 550-cubic-inch V-8 diesel engine good for 145 PTO horsepower. Three smaller models, the 354, 454, and 574, were also introduced in 1971. These utility tractors ranged from 33 to 53 PTO horsepower.

1972

In 1972, the 4156 and 656 joined the 66 Series and became the 4166 and 666. The 4166 shed the yellow-and-white color scheme used since the original 4100 and was painted red and white like the rest of the line. Both of these models were built until 1976.

The Steiger-Built Internationals

One of the simplest ways to add a big four-wheel-drive to the lineup was to purchase it from another company and rebadge it. Steiger had been providing such tractors for many years, and International turned to them in 1973 with the 4366. At some point, International actually was a part owner of Steiger, which made it more profitable for IHC to contract the company to build tractors.

The big diesel in the 4366 was an International 466-cubic-inch six-cylinder engine that was rated for 225 PTO horsepower. The model was an articulated four-wheel drive which was produced from 1973 to 1976.

In 1976, International put its 798-cubic-inch turbocharged V-8 diesel engine into a Steiger chassis and created the 4568. The big tractor weighed nearly 13 tons, was rated for 300 horsepower, and was produced only in 1976. In 1977, the two models were renamed the 4386 and the 4586, but the tractors were essentially the same units.

The last addition to the Steiger line came in 1979, when a more powerful version of the IHC 798-cubic-inch V-8 was used in the same basic frame. The tractor was the 4786, and it was rated for 350 horsepower. All of the Steiger tractors disappeared from the line in 1982, when International sold off its share in Steiger in an effort to recoup some of the company's staggering losses.

86 Series

Late in 1976, Harvester introduced the 86 Series, a line of new tractors that included the Hydro 86, 686, 886, Hydro 186, 986, 1086, 1486, 1586, and 4186. Two Steiger-built four-wheel-drives, the 4386 and 4586, were also offered with the new "86" model designated. International also introduced a new 284, a compact tractor built by Kimco, a company formed by a joint venture between International and Komatsu.

Control Cab

The two-wheel-drive 86 Series tractors featured the new Control Cab, which was moved a foot-and-a-half forward in the chassis. This reduced the amount of noise in the cab and improved the ride for the operator. The cab was part of Harvester's efforts to incorporate human factors in engineering. Human factors engineering first appeared in World War II, when developers discovered the importance of designing the machine to fit the operator. An example is non-standard locations of the switch that opens the airplane's canopy in an emergency. In a crash, the pilot will instinctively reach for the spot in which the lever was located in the plane they are most familiar with. If the plane they are flying has it located elsewhere, the time it takes to find it may make the difference between life and death. Engineers began to design the machine to meet the operator's needs, and improved equipment resulted.

By the 1970s, human factors engineering was a part of the farm tractors development, and the Control Cab resulted. The new cab gave the operator better vision to the front and easier entry to and exit from the cab, as well as the improved ride and quieter environment.

The amount of noise inside the cab became an issue with the tractor industry, and the tests at Nebraska began measuring how much noise reached the operator. Engineer Robert Oliver said, "What they were doing was trying to relate hearing loss to how long the farmer had to sit on the tractor, exposed to loud noises."

According to Oliver, quieter cabs led to the need for better instrumentation. The quieter cab isolated the farmer from the sounds of the machinery. It was harder to tell when something was going wrong, and the farmer had to be taught to watch gauges. "He couldn't tell if his chopper was plugging up, because he couldn't hear it. So, it was a whole learning experience getting the farmer to understand that he could run it with instruments. We had to start developing better tachometers and putting them in spots where they could see them better. It started the whole push on human factors engineering, so he could do all the things he needed to do inside that cab, and give him the comfort that he still had control."

The Need for Synchromesh

International took the lead on noise levels inside the cab, and the 86 Series was a successful venture. But they lagged behind in transmission technology. The industry was turning to synchronized transmissions that allowed shifting on the fly. Harvester was well aware of the lag, but a new transmission would require millions of dollars of retooling costs. The company was unwilling to sink the capital into the new line, and the 86 Series carried capable but aging transmissions.

The powershift transmission would come to IH, but it would be too late when it arrived. In the meantime, the company had another innovative tractor up its sleeve.

The New-For-1979 Models

In 1979, the 30- to 60-horsepower machines were improved, with some upgraded models and two new models. A more powerful Steiger-chassis machine was also added. The big news was the new 2+2 models, the four-wheel-drive 3388 and 3588. These two models were a whole new take on the four-wheel-drive tractor and were one of the company's most striking engineering advances of modern times.

The 2+2s gave the company another shot in the arm, and were a key part of a 1979 sales surge. A comparison of farm sales equipment found Deere & Company on top at $3.9 billion of annual sales, with IHC not far behind with $3.1 billion in sales. Massey-Ferguson was next with $2.5 billion, with Ford and Sperry New Holland the only others (just) over a billion dollars in 1979 sales. Bear in mind that these comparisons can be skewed significantly depending on how they are totaled (Do you include lawn tractors? Snow blowers?), so take the figures with a grain of salt.

The 2+2

Introduced as the International 3388 and 3588 in 1979, the 2+2 four-wheel-drive tractors were one of the company's last major engineering advances. The big tractors were articulated, as most of the big four-wheel-drives were, but the engine was placed well over the front wheels. By doing so, the big tractor had equal weight distribution over the front and rear wheels. The result was better traction without the addition of extra weights, and more even ground pressure. Also, the tractors could turn in a 15-foot, 9-inch radius, which was impressive for such big machines. The original Farmall, hailed for its quick turning, needed about the same amount of space to turn around.

The big tractor was more maneuverable than typical big four-wheel-drives, which were great for plowing but a bit clumsy for row crop work. The expensive tractor sold modestly, but received rave reviews.

Both 2+2 models were assembled mainly from existing components. The articulated joint was all new, of course, but most of the rear end came from the 86 Series tractors. The result was more economical production, and lowered design and tooling costs.

The 3388 used a 130-horsepower six-cylinder turbocharged diesel engine, while the 3588 used a bored version of the same engine to crank out 150 horsepower. While these numbers were impressive, the two were only about halfway up the IHC horsepower chart. The Steiger-built 4786 was powered by a turbocharged V-8 rated for 350 horsepower.

The tractors were equipped with both 540- and 1,000-rpm IPTOs, and the Torque Amplifier allowed 16 forward and 8 reverse speeds. The cabs of the 2+2s were insulated, air conditioned, heated, and sound deadened, keeping the operator comfortable in all conditions.

84 Series

After the major revamp of the larger tractors for 1977, International looked to bolster its small and mid-sized tractors in 1979. Four models, the 363, 464, 574, and 674, were replaced with new 84 Series tractors and two new models, the Hydro 84 and 784, were added to the line.

384

The 384 was rated as a 36 PTO-horsepower tractor and had a 154-cubic-inch diesel. The little tractor had eight forward speeds and served as a decent utility tractor, just slightly bigger than the larger garden tractors. The 384 was discontinued in 1982.

484

The new 484 used a 179-cubic-inch three-cylinder diesel engine that produced 42 PTO horsepower. The transmission had eight forward and four reverse speeds, and was equipped with a 540-rpm IPTO and Category I three-point hitch. Utility and row-crop models were sold.

584–784

The 584 through 784 ranged from 53 to 67 horsepower and used four-cylinder diesel engines. Category II three-point hitches were standard, and the models were available in utility, row-crop, low-profile, and all-wheel-drive versions. The two larger models had an optional torque amplifier, and the 784 had both 540- and 1,000-rpm IPTOs. The Hydro 84 was a 59-horsepower model equipped with hydrostatic drive.

844

The 844 was a new addition in 1980. It was a 73-horsepower model that used a 268-cubic-inch four-cylinder diesel. The tractor was equipped with dual IPTOs and a Torque Amplifier was optional.

Synchromesh!

The 84 Series tractors received synchromesh transmissions by 1983, and were produced through 1984 when the imminent merger of the International agricultural division and Case was announced. In addition, the 484, 584, and 684 were produced as Case-Internationals after the sale.

The new-for-1979 models performed well for International, and the year was a good one for the industry and IHC. It seemed that the company would be able to overcome the debts and uncertainty that shrouded the tractor industry. Despite such optimism and IHC's strong sales, the company had only about five years left.

Into the 1980s

As the 1980s dawned, International's struggles became cataclysmic. The debt load that hung around the company's neck was enormous, and the times were difficult ones for tractor manufacturers. Most farmers were struggling under debt loads similar (relatively) to that of International, due to the flood of money and available financing in the 1970s. In the early 1980s, farmers were refinancing or going belly up. The last thing on their minds was new tractors.

The early 1980s saw International introduce quite an assortment of new tractors. The 786 was added to the 86 line, and the 140 was replaced with the 274 Offset. The 140 was the long-lived descendant of the Model A, which was finally replaced. The 274 used a three-cylinder diesel engine and was built by Kimco, a division that was a joint venture between IHC and Komatsu.

The line was revised in 1982, with the new 30, 50, and 70 Series tractors as the stars. The 2+2 tractors were revamped, and the Cub derivatives received a make-over and a new model as well. Despite the introduction of the new tractors, net losses for 1982 were $1.6 billion.

50 Series

Tragically, Harvester introduced its new 50 Series tractors in 1982, with farmers struggling just to stay off the auction blocks. The 50 Series

Before International Harvester was purchased by Tenneco and became part of Case-IH in 1984, the company produced a line of high-tech tractors known as the 30 series and the 50 series (along with a very few 70 series articulated four-wheel-drives). This is a 3688.

tractors—the 5088, 5288, and 5488—featured the new synchronized transmission, but they were too late to avert the coming disaster. The new transmission consisted of a six-speed transmission coupled to a three-speed range transmission. The result was 18 forward and 6 reverse speeds. A differential lock could be engaged with a switch at the operator's heel, and the transmission could be power-shifted between first and second, third and fourth, or fifth and sixth.

The engine was a turbocharged six-cylinder diesel. The 5088 used a 436-cubic-inch powerplant rated for 136 PTO horsepower, while the 5288 and 5488 used a larger 466-cubic-inch engine that was rated for 163 and 187 PTO horsepower. A built-in ether injection system was supplied for cold-weather starting, and the engine and drivetrain warranty was for three years or 2,500 hours of use.

Modern tractors were becoming much more powerful for their size. The 5488 weighed about 15,000 pounds in two-wheel-drive form, and produced 187 PTO horsepower. Those figures compare quite favorably to the 65 horsepower of the 10,000-pound Super WD-9 back in 1954.

The 50 Series tractors used three hydraulic pumps, like earlier models, but the pumps were more powerful and mounted outside the transmission housing for easy servicing. All-wheel-drive was an option on all three models.

One of the features of the 50 Series was the reverse-flow radiator, a simple idea that had been around since the optional reverse fan on the Letter Series tractors. By drawing air from above the hood and pushing it out through the radiator, dust and chaff were blown off the radiator rather than sucked in. Also, the hot air blew away from the cab, giving the air conditioner a bit of a break.

The cab had become an amazing place by the 1980s, and the 50 Series was no exception. Heating and air conditioning were standard, and the seat and steering wheel were adjustable. The seat was suspended with an air and hydraulic fluid system, and featured arm rests and lumbar support. The International cab was one of the quietest in the industry, and had doors on both sides for easy access. A computer display gave digital readouts of engine rpm, ground speed, exhaust gas temperature, and PTO speed. The cab had integral roll-over protection (ROPS).

30 Series

The 30 Series tractors were introduced in 1982. The 90-horsepower 3288 and 110-horsepower 3688 were introduced in 1982 along with the new 50 Series machines. These two tractors had a bit less power and fewer features than the 50 Series. The 3688 used a normally aspirated version of the 436-cubic-inch engine in the 5088, while the 3288 used a smaller 358-cubic-inch normally aspirated diesel powerplant.

The 1983 additions to the line were the 112-horsepower 3488 Hydro and the 80-horsepower 3088. All but the hydrostatic model used the same synchromesh transmission found in the 50 Series machines, with 16 forward and 8 reverse speeds. The forward-flow radiator was standard, as was the 540/1,000-rpm IPTOs, differential lock, and the heated and air-conditioned cab. The 30 Series had fewer bells and whistles as standard equipment, with the data center and plush interior optional on most of the line.

Modern International Tractor Production

Does not include tractors shown in previous charts.

Model Name	PTO Hp	Production Years
4100		1965–68 856†
4156		1969–70
4166		1972–76
4186		1976–78

Steiger-Built Chassis

4366	225e	1973–76
4386	230e	1977–81
4568	300e	1976
4586	300e	1977–81
4786	350e	1979–81

I-606		1963–67
I-706		1963–67
F-706		1963–67
I-806		1963–67
F-806		1963–67
F-404		1963–67
I-404		1963–68
F-504		1963–68
I-504		1963–68
I-544		1968–73
F-544		1968–73
I-424		1964–67
I-444		1967–71
I-454		1970–73
464	44	1974–77
574	52	1970–78
674	61	1974–77
F-656		1965–72
I-656		1965–73
666		1972–76
I-664		1972–73
674		1973–77
Hydro 70	69	1974–75
Hydro 100	103	1974–75
1566	160	1975–76
364	36	1976–77

86 Series

Hydro 86	69	1977–80

686	66	1977–80
886	85	1977–81
Hydro 186	104	1977–81
986	105	1977–81
1086	130	1977–81
1486	145	1977–81
1586	160	1977–81
4186	175	1977–78
786	80	1981

84 Series

284		–86*
384	36	1978–80
484	42	1979–85*
584	52	1979–85*
Hydro 84	56	1979–84
684	62	1979–85*
784	65	1979–84
884	72	1980–85*

2+2

3388	130	1979–81
3588	150	1979–81
3788	170	1980–81

30 Series

3288	90	1982–84
3688	110	1982–84
3088	80	1983–84
3488 Hydro	112	1983–84

50 Series

5088	135	1982–84
5288	160	1982–84
5488	185	1982–84

60 Series (2+2)

6388	130	1982–84
6588	150	1982–84
6788	170	1983–84

These tractors continued to be produced by Case-IH after the merger.
Source: †McCormick/IHC Archives, "Tractor Production Schedule"
Production figures are number of serial numbers and do not include the last year of production. These figures are only rough approximates and are rounded off.

60 Series

The 60 Series was an upgrade of the 2+2 models and became the 6388 and 6588 that appeared in 1982, with a new 6788 added to the line in 1983. The two larger models received a somewhat larger engine, with 30 additional cubic inches bringing displacement to 466. They also received the new synchromesh transmission, allowing power-shifting at speed. The 6388 and 6588 had 16 forward and 8 reverse speeds, while the 6788 had 12 forward and 6 reverse speeds. These were the final and finest versions of the 2+2, as these tractors were dropped from the line after the merger, making 1984 the last year 2+2 tractors were built.

70 Series

A new 70 series of big four-wheel-drive tractors was supposed to be introduced in 1985, but never made it beyond the farm shows in September 1984. The tractors were in the 200-PTO-horsepower class, probably as replacements for the Steiger-built machines that were dropped in 1982. The sale of the International agricultural division was announced in November of 1984, and the big 70 Series machines were never produced.

Fall From Grace

Several factors contributed to IHC's fall. The first was the aforementioned debt load. Even during the boom of the 1970s, the company could not escape this burden. Management is blamed for making several key mistakes that led to the company's downfall, but it may be that the debt load was such that the company simply could not survive the times.

Another one of International's problems was with organized labor. Although the company had a long history of treating employees well, it had an equally lengthy record of struggles with unions. The McCormicks refused to bargain with organized labor, and that tone continued to the bitter end. Although International's employees were treated relatively well, management never seemed to master the art of dealing with organized labor.

In 1979, after the company experienced record-breaking profits, management badly misjudged organized labor. Management pushed to bring wages and benefits down, and labor struck. International leaders refused to give and fought a losing battle for nearly six months. In the same time span, the economy swung toward recession, and the company lost money that it could not afford to throw away. When the workers returned, International leaders misjudged the level of market demand built up during the strike, and the company overproduced tractors for a sagging market. By the time the 50 Series was introduced—the tractor that was supposed to save the farm equipment division—demand was almost nil and the company was incredibly overstocked with suddenly outdated machines.

The company's administration had been investigating the possibility of filing for chapter eleven bankruptcy protection since 1981. The move would have been unprecedented; no company of International's size had ever filed for chapter eleven. The banks were hounding the company, and the company was finding it difficult to obtain loans while losing millions each year. International began slashing costs wherever possible and, by the mid-eighties, rumors of a sale were circulating in the agricultural division. Harold Schram recalls these days as the only unpleasant time during his more than 25 years with the company.

"There were constant rumors about who was coming through to look at the ag equipment division. We were told in mid-November of 1984 that there wasn't any substance to the rumors about the sale to anybody and, of course, it was rumored to be Tenneco. The papers here were even carrying articles that said the president of our division had been seen coming out of the Tenneco corporate office in Houston, Texas. About two weeks after we were told there was no substance to the rumors, the sale was announced."

In 1984, the farm equipment division was sold to Tenneco, the owners of J. I. Case. Case and International's tractor division were merged, and the tractors that ensued were known as Case-Internationals. Veterans of IHC and those close to the company say that the merger was not good for the International Harvester agricultural division. Plants were closed and workers from all venues were sent home with pink slips. Most of the IHC models were discontinued, and almost all were replaced in the first few years of production. The Hinsdale Farm, the site of endless hours of testing of all kinds of IHC product through the years, from tractor to implements to milking machines, was divided up and slowly sold off for development.

The CNH Merger

The Case Corporation was merged with New Holland in 1999. The merger reduced the number of major players in the worldwide tractor manufacturing business to four: John Deere, AGCO, Deutz-Farr, and

CNH. At the turn of the 19th century, the agricultural implement industry was conglomerated to form the International Harvester Company. In recent history, the tractor industry is going through a similar transition and has become increasingly concentrated.

The merger of New Holland and the Case Corporation created a $12 billion company called CNH Global N.V. which has more than 12,000 dealers in 160 countries. This giant corporation has taken dozens of brand names under it's wing, including Braud, Case, Claeys, Fiat, Flexicoil, Ford, International Harvester, New Holland, Steyr, and many others. The primary brands are New Holland and Case-IH.

As was done in 1902, those two companies united under CNH are attempting to operate as independent entities. Case-IH and New Holland tractors and equipment have separate brand names and dealership networks. The two lines share plenty of products and parts, and some of the lines are merely the same machines painted in their own colors.

When the International Harvester company was formed, the merger attracted the attention of the U.S. Justice Department and, after a decade-long court case, IH was required to sell off a number of properties. The CNH merger received a close look from the U.S. Justice Department and also attracted the attention of the European Union, as New Holland is a European-based company.

The wheels of justice move more quickly today, and the results of these investigations required the companies to sell off a number of product lines. In order for the merger to pass muster, New Holland had to dump their Kansas-based Hay & Forage Industries, the Versatile line of four-wheel-drive tractors and the Genesis line of two-wheel-drive tractors, along with the plant in Winnipeg, Canada that built these two tractor lines. New Holland also was required to sell its Fermec backhoe/loader, industrial, CX and Mxc tractor lines, along with the plants that build them in Manchester and Doncaster, England. New Holland also had to sell an Italian-based combine harvester line.

These sales comprise only 3 percent of the total product line offered by the companies. In a somewhat ironic twist, the sale of the Doncaster plant and the line of C, CX and MC, and MTX tractors resulted in more of the IHC heritage coming back. The plants and the line of tractors were purchased by an Italian equipment company, Landini, which is part of the European ag equipment conglomerate, Argo. The new company is building a line of McCormick tractors ranging from 58 to 280 horsepower, and has sales offices in North America, Europe, and Asia.

The McCormick tractors are mainly intended for European sales, but had 112 U.S. dealers in 2002 and had brought over about 800 tractors. The company has a company headquarters in Pella, Iowa, and has recently introduced a new line of 100- to 160-horsepower MTX Series tractors.

Another interesting result of the merger has been the resurrection of the Farmall brand. Case-IH surveyed their customers, and found that the Farmall brand had the strongest recognition of any of their historic names. As a result, the line of 14- to 47-horsepower compact tractors originally built by New Holland, and known as the "Boomer" tractors, will appear in Case-IH dealerships as the Farmall D- and DX-series tractors. CNH promises to re-tool the controls and look of the machines, but underneath the skin, they will closely resemble the Boomer line.

Farmall is back as a brand satisfying a market segment, part of a global corporation that is one of the world's largest companies. These giant corporations understand the importance of branding and marketing and take advantage of opportunities to turn their legacy into sales. Case-IH sponsored a tour by country music star Craig Morgan, who had a hit song out entitled, "International Harvester."

Jim Walker, the vice president of Case IH North American Agricultural Business described the link between Morgan's music and IHC as a natural one.

"The same people who are Craig Morgan fans ought to be Case IH fans," he said. "If they like being in a concert seat, they should try the seat of a red Farmall tractor."

The marketing song might be the same, but IH employees speak fondly of the company as a place where your co-workers were like family, the management was team-oriented, and your future was assured.

Long-time IHC engineer Gordon Hershman's comments are typical of those who spent a portion of lives with the company. "I love Harvester. They were a tremendous company. I always feel bad that we lost it the way that we did."

Harold Schram, another long-time engineer recalls his days with the company as some of the best of his life. "If you worked for Harvester, you were part of the family and you felt that way. There was a lot of loyalty both ways—employees to the company and company to the employees... It was more than just a job to us—it was a way of life."

That way of life is gone, but the tractors live on.

Appendix

Recommended Reading

Gray, R. B. *The Agricultural Tractor: 1855-1950*. U.S. Department of Agriculture, 1954.
Although a bit dry and sometimes brief on specific information on tractors, this book is a treasure trove of information about how tractor technology changed and what effects it had on the farm.

Larsen, Lester. *Farm Tractors 1950–1975*. American Society of Agricultural Engineers, 1981.
This book contains lots of interesting information on how the tractor progressed from 1950 to 1975. There's a partial listing of models for each year, and brief discussions of significant new technologies such as roll-over protection, turbochargers, and the development of diesel and LPG engines.

Gay, Larry. *Farm Tractors 1975–1995*. American Society of Agricultural Engineers, 1995.
The only source of its kind for a complete list of modern tractors. The charts in the back showing the progression of models and the complexity of mergers and corporate changes are worth the price of admission alone.

Marsh, Barbara. *A Corporate Tragedy*, Doubleday & Company, 1985.
If you want to know how the International Harvester Company management evolved and why the agricultural division was sold to Tenneco, Barbara Marsh's book is the one to have. It is well-researched, and Marsh tells an interesting story well, with lots of interviews with ex-IHC employees. Great stuff. Tragically, the book is out of print.

McCormick, Cyrus. *The Century of the Reaper*, Houghton Mifflin Company, 1931.
If *The Agricultural Tractor* is dry, this book is the Sahara desert. But there are some great passages in here and the early history of the company is told with an occasionally colorful but always slanted voice. This book is also out of print and quite rare; snap one up if you find one for a reasonable price.

Wendel, C. H. *150 Years of International Harvester.* Crestline, 1981.
This hardcover Crestline has been the source for IHC enthusiasts for many years, and the book covers the company's products in unparalleled detail. Wendel did an incredible job gathering this information, and there's lots of good stuff inside, but double-check the data if possible. No matter, where else can you find a book with listings for everything from disk harrows to cream separators? A must.

Williams, Robert C. Fordson, Farmall, and Poppin' Johnny, University of Illinois Press, 1987.
This is a graduate thesis that examines how and why the farm tractor changed the farm. This is good stuff, and a serious study of the tractor and the farm. Williams filled out what R. B. Gray started with *The Agricultural Tractor,* and improved on the original.

Leffingwell, Randy. *Farmall: Eight Decades of Innovation,* MBI, 2005.
You can count on Leffingwell to do extensive research on any of his books, and to uncover new revelations about the brand. He came through again with this book, and it is a must-have for any serious IH history buff.

Fay, Guy. *International Harvester Experimental and Prototype Tractors,* Motorbooks International, 1997.
Guy Fay was one of the curators at the IHC company archives housed at the University of Wisconsin, and his research has enlightened IH enthusiasts in both the printed word and online. Any of his book's are thoroughly researched, and this one is a particularly interesting assembly of information about the rarest IHC tractors.

Baumheckel, Ralph. *International Harvester Farm Equipment Product History 1831–1985,* ASAE, 1997.
Well-researched exhaustive product history of the company. A must.

Fay, Guy.
The original series guides give you all the details about parts and options that are correct for your Farmall. A must for any serious restorer.

Klancher, Lee. *Farmall: The Golden Age 1924–54,* Motorbooks, 2002.
A recounting of the effec the tractor had on society from 1924 to 1954, illustrated with original color images from the IHC archives.

Clubs, Mags, and More

International-Harvester Collectors
Association
648 N. Northwest Hwy.
Park Ridge, IL 60068
www.nationalihcollectors.com/

Harvester Highlights
18324 Monroe Road 1073
Madison, MO 65263
660/291-8742

Red Power Magazine
PO Box 245
Ida Grove, IA 51445
www.redpowermagazine.com

Antique Power Magazine
P.O. Box 1000
Westerville, OH 43081-7000
888/760-8108
www.antiquepower.com

Belt Pulley Magazine
PO Box 58
Jefferson, WI 53549
920/674-9732
www.beltpulley.com

Early Day Engine and Tractor
Association
1537 Weekend Villa Rd
Ramona, CA 92065
760/789-3402
www.edgeta.org

Engineers and Engines
Outrange Farm Publications
PO Box 10, Bethlehem, MD 21609-0010
410/673-2414
www.eandemagazine.com

The Hook (Pullers' Magazine)
P.O. Box 937
Powell, OH 43065-0937
614/848-5038
www.hookmagazine.com

Antique Tractor Internet Service
www.atis.net
Enthusiast site to chat and learn about
tractors. The first and one of the best.

Yesterday's Tractors
www.yesterdaystractors.com/
Enthusiast site.

Hotline Guides
www.hotlineguides.com/
Used tractor values and production

TractorData
www.tractordata.com/
Lots of great information about nearly
every tractor ever built

Case-IH
www.caseih.com

Acknowledgments

The *Farmall Dynasty* is an updated version of my second book, *International Harvester Photographic History.* That original book could not have been published without help from Guy Fay and Cindy Knight for their help doing research. The time donated by engineers Harold Schram and Robert Oliver was also a tremendous help. A number of people reviewed drafts of the book and provided valuable input: these include Jim Becker, the late Scott Satterlund, Steve Offiler, and Matt Laubach. Zack Miller, Michael Dapper, Anne McKenna, Jack Savage, and Katie Finney helped make it a book at Motorbooks.

The current edition was proofread by Peter Bodensteiner and the design was put together by Cory Anderson. Thanks to both!

About the Author

Born and raised on the banks of the Brill River in northern Wisconsin, books are a life-long passion for author Lee Klancher. Lee has authored eight books, including *Farmall: The Golden Age 1924-54, The Tractor in the Pasture,* and *How To Build Your Dream Garage* (all of which can be found on Amazon.com). His other work includes photographing farm tractor calendars and nearly a hundred magazine articles covering everything from a tour of post-Katrina New Orleans to photographing the Bolivian national phenomena known as Caravana. Lee's work appears regularly in motorcycle and travel magazines.

Lee's other interests include off-road motorcycling, racquetball, mountain biking, dive bars, college hockey, and the Green Bay Packers.

You can find samples of Lee's work, book reviews, interviews, and contact information at www.leeklancher.com.

Made in the USA
Lexington, KY
07 December 2011

This book gave our church the confidence to step out and obey Jesus' standing orders to heal the sick. We've seen much more healing in our community after applying this teaching. — *Josh & Jessica Morgan, House Church Pastors, Greenville, NC*

Do you ever get asked to pray for someone who is sick? Of course you do, and of course you want to see them healed. Unfortunately, there has been so much confusion around how to effectively pray for the sick that the average Christian simply does not know what to do. *Healing the Sick* by Alex and Hannah Absalom is a great book for equipping every Christian to do what the book's title says, heal the sick. It is theologically sound, easy to read, and practical in application. It is a must read! — *David Patterson, Lead Pastor, River Cross Church, A Southern Baptist Church, Gainesville, FL*

I have been looking for a book like this for years! This resource is holistic; theological yet accessible, spiritual yet practical, and profoundly captures the heart of the Father, the ministry of the Son, and the power of the Spirit. — *Chris Brister, Lead Pastor, Union Church, Auburn, AL*

As someone who has a great desire to heal the sick and train my church to do the same (but who had zero training or experience in how to do it), this book is an incredibly helpful guide. Practical, applicable, humble, biblical. I'll be taking my whole church through this book. — *Josh Wagner, Pastor at United City Church, Indianapolis, IN*

Everyday healing for everyday Christians. Alex and Hannah Absalom have provided an incredibly detailed and concise resource full of Scripture, testimonials, realism, and practical next steps. It will make you want to love people by praying for their healing. — *Jerry Fourroux, Jr., Pastor, Lycoming Centre Presbyterian Church, Cogan Station, PA*

After reading this book, I feel invited and excited to see Christians and the church walk with Jesus in such a way that we see God's power and love flow to those who suffer and the gospel and glory of God breakthrough to more and more people. It's a well written, practical, gracious, humble and theologically solid book that will invite you into deeper intimacy with Jesus to see His love and power experienced by more and more people. — *Jarrod Hawthorne, Pastor, The Crossing Church, Monroe, LA*

I love how Alex and Hannah are able to take spiritual concepts that can seem overwhelming and make them reachable and practical for all of us! *Healing the Sick* is just that kind of book. When you finish each chapter, you'll find yourself saying, 'Well, now I can do that!' And you're right - you can! Jesus-centered, theologically sound, and incredibly practical. Buy it, read it, and reference it again and again. — *Brian Kannel, Lead Pastor, York Alliance Church*

Alex and Hannah's work serves to locate the disciple of Jesus firmly within the story of God's in-breaking Kingdom. It is an important reminder that, as disciples of Jesus, we are called to participation in the redemption and restoration of all things by the power of God's Holy Spirit. Life in the Kingdom of God is an adventure and Alex and Hannah remind us that where Jesus goes, so too go life and goodness and healing! — *Matt Moore, Lead Pastor, City Collective Church, Chattanooga, TN*

Alex and Hannah provide a winsome, practically view of the ministry of divine healing. It is filled with wisdom, heart, transparency, and real life stories that will lift your heart and fill you with faith that our God is a healer! My new favorite on the subject. — *Dr. Kwesi Kamau, Impact Discipleship*

HEALING THE SICK

BIBLICAL AND PRACTICAL WISDOM FOR
HEALING THE SICK IN NATURALLY
SUPERNATURAL WAYS

ALEX AND HANNAH ABSALOM

Dandelion
Resourcing
dandelionresourcing.com

For our favourite children — Joel, Samuel, and Isaac

CONTENTS

INTRODUCTION

Welcome To The Naturally Supernatural Series!

We are a family of missionaries - Mum, Dad, and three sons - who in 2007 were called to move from England to the United States, back when our boys were small and Alex had more hair!

Having now lived in three very different parts of the country, we can say that we love this calling! We have had the privilege of speaking to tens of thousands of Christians, coaching many hundreds of leaders personally, consulting with dozens of churches and denominations, as well as always being fully involved in local church leadership - ranging from megachurches to church planting.

During this time, it has been exciting to see the wider church in the US develop a stronger emphasis upon discipleship, particularly a drive towards creating cultures that intentionally form disciple-making disciples of Jesus. There is a deeper recognition that discipleship must be built around life imitation, rather than simply reducing it to information transfer.

Likewise, we have also observed an increased focus on equipping believers to go with the Gospel. More people are being actively encouraged to carry the words and works of Jesus into their neighborhoods, communities, workplaces, and schools. While great Sunday morning church services are wonderful, growing numbers of leaders are realizing that simultaneously we must also encourage new expressions of church to form throughout the week, wherever the lost are already gathering and being discipled.

THE MISSING PIECE

Yet there is still a piece missing alongside disciple-making and living on mission, which for us is summed up by the phrase 'naturally supernatural'. **As disciples of Jesus, we need to be people who operate in the power and the authority of the Holy Spirit.**

Sometimes this will mean opening our mouths and sharing the Good News of the Gospel. Other times it will be modeling the servant leadership that Jesus calls us to embody. And on other occasions we will make use of the gifts of the Spirit, bringing healing, deliverance, prophecies and the like as part of the expansion of God's heavenly Kingdom here on earth.

In all of these things, we will be empowered to exhibit the fruit of the Spirit to all who encounter us, because if we do not have love, we are nothing (1 Corinthians 13:2-3).

To be naturally supernatural means that we demonstrate these more supernatural marks of the Spirit's empowering throughout our lives - to the extent that it simply seems natural and normal to anyone observing. And while we know that these things can only come from God's Spirit, at the same time they should be normal and normative, rather than strange and random.

One of our surprises living in the US has been the apathy in many Christians towards this life in the Spirit, particularly to the more seemingly supernatural gifts. (Of course, every gift requires the anointing of the Spirit to do well, but we recognize that there are some that are genuinely impossible for us to produce simply by our hard work.)

Perhaps it is the relative ease of calling yourself a Christian in the United States when compared to most other Western nations, let alone the rest of the world. If deciding to follow Christ proves to be a culturally comfortable choice, then it makes sense that many will not feel a daily need to press into the Spirit's authority and continual empowering.

In some cases this is due to open hostility towards the gifts being active today. Cessationism (the belief that the gifts of the Spirit died out once the Bible was written) has had an especially warm reception amongst some American denominations. It is a classic example of a theology being built in an attempt to justify an experience (in this case, a lack of supernatural activity), which tragically has resulted in countless Christians being taught that the gifts of the Spirit 'are not for today'. **Of course, the same challenge is likewise there for those whose worldview does allow for naturally supernatural ministry: we must first start from Scripture, and then allow that to interpret experience, rather than the other way around.**

For others, the issue is not so much theological as practical - just how do we operate in these things while remaining a people deeply shaped by Scripture? And where can we find clear and wise practices to learn from? This is the imitation paradox: many people we speak with have never seen naturally supernatural living modeled in ways that are accessible and desirable, and so, in turn, they don't know how to model it to those they lead.

OUR BACKGROUNDS: EVANGELICAL AND NATURALLY SUPERNATURAL

Of course, there are some wonderful churches and ministries who beautifully model supernatural living. We have been fortunate to experience many such communities, and have benefitted from their stories, experiences, and teachings. However, many who come from evangelical backgrounds often don't feel comfortable in such environments, and much is lost in translation between the two tribes.

Our own stories of encountering life in the Spirit (which we will tell in greater depth in our book *Being Filled with the Holy Spirit*, which is part of this series) have always been rooted in a strong evangelical theology. The centrality of the cross of Christ and His atoning work there, belief in the literal resurrection, a high view of Scripture, the urgency of sharing the Good News of Jesus, and an understanding that following Jesus has to be lived out in the world, are foundational planks of our faith.

To pursue the present day activity of the Holy Spirit through the gifts of the Spirit requires zero diminishment of those core beliefs. We would argue that a naturally supernatural lifestyle is one that is far more faithful to the clear commands of Christ, the example of the early church in Acts, and the teaching of Paul. It is actually MORE biblical to be naturally supernatural in your beliefs and practice!

We have been in full-time church leadership since 1994, and throughout that time we have pursued both a deeper commitment to Scripture and a simultaneous deepening of our experience of the Spirit's empowering. One feeds the other, and thus we are drawn closer to Christ and, hopefully, to greater conformity to Him.

WHY THIS SERIES

Over recent years we have found ourselves teaching more and more on living a naturally supernatural lifestyle, both at conferences and with individuals and small groups. Friends began to encourage us to write on this area, but, to be honest, we didn't want to do so unless Jesus really made this clear to us.

One day, as Alex was spending time with the Lord, he unexpectedly saw a clear vision of a series of shorter books that tackled different topics, all under the banner of 'Naturally Supernatural'. The mandate was to produce something that would equip the church, combining strong Scriptural teaching with field-tested practical 'how tos' for implementation, mixed in with stories, activation exercises, study guides, passages to memorize, Q&As, prayers, and anything that would help individuals, groups, and churches to step into these areas with these books as a guide. For us, this book series does indeed feel like a call from God.

Topics we cover in the series include the prophetic gifts, healing, deliverance, building a Kingdom theology, being filled with the Holy Spirit, tongues, and becoming a naturally supernatural missionary.

WHY THIS BOOK

Can you remember that last time you had a thumping headache, a strained muscle, or an upset tummy? Such experiences can quickly become all-consuming, even if the sensible part of your mind knows that they will soon go away by themselves. And how much more do you find yourself overtaken when the illness is a major one, especially when someone very close and dear is the one who is suffering.

As people made in the image of God, we are not designed to have lives marked by sickness. However, since the Fall this has been part of our common experience. Humanity has done an incredible job coming up with all the rich range of treatments that modern medicine provides, but it can't cure everything, even in the unlikely scenario that we have full and instant access to all of its options.

It is into the strife of sickness that God loves to insert Himself, *"For I am the Lord, who heals you"* (Exodus 15:26). This revelation is made most clear through the ministry of Jesus. *"Jesus went throughout Galilee, teaching in their synagogues, proclaiming the good news of the kingdom, and healing every disease and sickness among the people."* (Matthew 4:23)

Yet this was not a raw expression of divine power that left us simply admiring God's power from a distance. Instead, all disciples of Jesus are commanded and equipped to carry on this ministry of healing, as part of our proclamation and demonstration of the breaking in of God's Kingly rule into the world around us today. *"When Jesus had called the Twelve together, he gave them power and authority to drive out all demons and to cure diseases, and he sent them out to proclaim the kingdom of God and to heal the sick."* (Luke 9:1-2)

This calling was not simply for the original disciples, but instead is a mandate for all believers in all places at all times. *"Philip went down to a city in Samaria and proclaimed the Messiah there. When the crowds heard Philip and saw the signs he performed, they all paid close attention to what he said. For with shrieks, impure spirits came out of many, and many who were paralyzed or lame were healed."* (Acts 8:5-7)

Healing the sick is the clear and unmistakable command from Jesus to each one of us who believes. *"As you go, proclaim this message: 'The kingdom of heaven has come near.' Heal the sick, raise the dead, cleanse those who have leprosy, drive out demons. Freely you have received; freely give."* (Matthew 10:7-8)

Our bedrock belief, which is based on the extensive teaching of the Bible, and our own experience from over 25 years of ministry, is that Jesus heals today! And He wants to use regular people like you and me to carry on doing so into the future!

In this book we will show you the extensive Scriptural instruction on why and how to heal the sick in naturally supernatural ways. As this is outside of the daily experience and theology of many, you will gain great clarity about what the Bible actually teaches and, even more importantly, what Jesus expects of you.

You will also discover a clear five-step model for ministering healing, to help you operate in a gracious, low-hype, loving manner. You will find this to be enormously helpful if you're hoping to develop a healing culture in your church or group. Along the way we will offer practical tips, give you shareable training, and reveal pitfalls to avoid, so that you can step into healing the sick in ways that are positive and life-giving to others.

Alongside this coaching there is also an extensive treatment of what to do when the healing doesn't come. **It is important to build a robust theology that neither ignores the reality that not everyone will be healed in this life, nor centers so much on those frustrations and heartaches that we lose faith to minister healing elsewhere.**

You will see that the book is divided into two sections: core teaching and next steps. Our desire is to give you a clear grounding in Scriptural principles and practical teaching on healing the sick (core teaching), after which are added a wide variety of practical on-ramps, designed to suit different personalities and spiritualities (next steps). Have fun playing with these (yes, you can have fun when experimenting with becoming more naturally supernatural!), and allow Jesus to show you where you can grow and develop further.

The goal throughout is to help you take specific, tangible next steps, and then to be able to help other people do likewise. This will mean taking risks - but it is a lifestyle that Jesus commends!

"When you enter a town and are welcomed, eat what is offered to you. Heal the sick who are there and tell them, 'The kingdom of God has come near to you.'"
Luke 10:8–9

PART I

CORE TEACHING

The main part of this book is focused on giving you a strong foundation from the Scriptures for healing the sick. You'll also discover proven practical training that equips you in the actual mechanics of how to heal, in ways that are bold yet wise.

Remember that Jesus warns against being the one who hears His words but fails to put them into practice. Therefore, we will focus unashamedly on your next steps, so putting what you read into practice is woven throughout!

This section will come to you in 6 chapters:

1. Our Healing Problem - Setting out some core understanding around what it means to heal the sick, and responding to some of the more common objections or excuses for not pursuing this gift.

2. A Biblical Theology of Healing - Showing how the gift of healing is deeply rooted in the ample teaching of the New

Testament, and then drawing out some of the theological principles that need to guide our everyday practice.

3. Principles for Praying - Identifying why a right understanding of authority is vital to living a naturally supernatural lifestyle, and assessing what Jesus actually means when He repeatedly ties healing to faith.

4. A Practical Guide for Healing the Sick - Sharing practical and proven guidelines for healing the sick, centered on our version of the 5 Step Model, and adding in lots of tips and cautions from our mistakes (and successes)!

5. When the Healing Doesn't Come - Tackling the greatest fear ("What if nothing happens?") by offering both pastoral and theological responses, and discussing why healing doesn't always happen and how to still persevere.

6. A Final Encouragement - Motivating you to put into practice what you're reading, since the goal of this book is that you grow in faith and persistence in healing the sick!

OUR HEALING PROBLEM

Some time ago we were invited by a church network to be part of a leadership prayer gathering to seek breakthrough for three key ladies who were battling life-threatening illnesses. After a time of worship and scriptural exhortation, the 40 or so present turned to prayer. The three seeking healing experienced extended intercession, the laying-on of hands, and anointing with oil. At the end of the time together, each of them expressed how much they felt truly loved and encouraged, in what was a faith-filled and highly caring environment.

Following that session, one of the three experienced a remarkable healing from Multiple Sclerosis. Her first clue was that her high levels of pain, which had been a daily constant for over three years, completely disappeared during our time of prayer. In the months that followed she closely observed her body, and did not have a single episode or symptom of MS!

At first the lady and her husband kept the healing to themselves, as it seemed too good to be true. After going through a battery of tests, her amazed doctors could offer no medical explanation for the MS not only going into remission, but also completely disappearing from her body. Eventually the joyous couple began to share and celebrate the news with family, friends, and their church. They commented, "It's been a world of emotions for us. The highlight was telling our daughters. When I asked my five-year-old what she was thinking, she yelled 'I love God. He fixed my Mommy!' So glad our girls will grow up knowing that their God is good and caring."

Truly, that is a wonderful story, which brings great glory to Jesus!

In the same time frame as that remarkable healing transformed one household, one of the other ladies who sought prayer was informed that her late stage cancer had aggressively returned and had spread widely, debilitating her body to such an extent that eventually she was moved into hospice care. Tragically, she passed away about 5 months after the leadership gathering.

Finally, the third person who sought prayer continues to fight her chronic illness, receiving medical help to mixed effect. She has been remarkable throughout her long fight by maintaining a sweet spirit and unwavering faith. Her desire is to pray for and bless others more than think about herself, and she has been a huge inspiration to many. And yet, she continues to wait for her physical healing.

Three remarkable women, each of whom loved the Lord, modeled tremendous fruitfulness in the Kingdom, and experienced prayer for healing from faith-filled leaders and elders. Yet three very different outcomes occurred - a miraculous healing, a death, and a continuing fight against chronic illness.

STRONG EMOTIONS

As you read these stories, you are probably already thinking about people you know who have faced major health battles. And like the women we prayed over, some of them experienced clear and remarkable supernatural healing, others were healed through the gift of modern medicine, a few died despite much prayer, and some continue to battle disease to this day.

Imagine if we could gather together and share these stories with one other.

The healings would be cause for great joy and praise, as we thanked God for His transforming intervention in the life of a person previously held captive by sickness. Yet alongside those moments, it would take a heart of stone not to be deeply moved for those where healing did not occur. Indeed, we would perhaps find strong emotions welling up within us - we should want to shout, **"This is NOT right! This is NOT how things should be! We hate sickness and cancer and death, and we want to see the presence today of Jesus' coming Kingdom, where such things will be NO MORE!"**

And while some might find such strong emotions unsettling, to us they are entirely appropriate responses as disciples of Jesus. We aren't meant to settle for some sort of compromised 'peace deal' with the devil, where we begin to accept that sickness is just one of those things with which we have to live, and for which we can have no spiritual response.

Our terms of living are to be set by Scripture, and the inspirational calling Jesus sets before us of joining Him in *"making all things new"*. Our theological worldview must not be determined by our experiences of disappointment, or the realm of hopelessness into which our enemy seeks to entrap us, but instead by the greater truths that Jesus reveals.

The good news is that we can indeed have a spiritual response!

Jesus not only healed the sick, but He also commissioned all of His disciples to heal the sick.

Which means you, and us, and every other Christian you know or have ever met!

So central was this task in the mind of Jesus that He inextricably tied healing the sick to our proclamation of the Gospel. For instance, in Luke 9:1-2 we read that Jesus gave His disciples *"power and authority to drive out all demons and to cure diseases, and he sent them out to proclaim the kingdom of God and to heal the sick."*

Yet how do we do this in practice? What is the solid biblical basis for such an approach? How can we ensure that we operate in a way that honors Christ in both word and deed? Perhaps you've seen abuses of this gift, so is it possible not to fall into those traps? How do we balance walking by faith with being pastorally sensitive? And what do we say when the healing doesn't take place either immediately or at all?

The goal of this book is to equip you to face these questions related to healing the sick, and to give you down-to-earth, proven guidance in the actual practicalities of how to do it.

We believe that healing the sick is a clear call that Jesus gives to every single one of His followers. However, a lack of healthy modeling, understanding, and clear action steps often impedes our obedience. Therefore, you will find in this book inspirational stories, clear biblical teaching, and wise best practices that you can start to implement today.

IS HEALING THE SICK REALLY FOR EVERY FOLLOWER OF JESUS?

Yes!

Everyone who follows Jesus is called to intentionally participate in healing the sick.

"As you go, proclaim this message: 'The kingdom of heaven has come near.' Heal the sick, raise the dead, cleanse those who have leprosy, drive out demons. Freely you have received; freely give." (Matthew 10:7-8)

You do not need to be a spiritual giant to do this! We've often seen people who are young in faith or young in years powerfully minister healing.

Some are given a sustained and ongoing ministry in healing, while for others it will simply occur as the need arises. Like any spiritual ability, you can grow in confidence, maturity, and fruitfulness in healing the sick.

Our job is simply to love the one needing healing, and to trust that Jesus will bring them health, wholeness, and life.

In fact, we believe that whenever we pray for healing with someone, the most important outcome is that they know and feel afresh the Father's love for them. That's even more important than the actual healing! (More on that in a moment…)

The Bible seems to show that faith is an important ingredient in the mix (even if that faith feels small and weak), along with an openness to God's healing power, and a willingness to persevere where that is needed. None of those ingredients should be a barrier to entry for any Jesus follower.

Anyone who has committed their life to Christ is called to heal the sick, break the hold of the enemy, and proclaim that the Kingdom of God has come in Jesus. These are non-negotiables, and just because many Christians choose to ignore them, that doesn't in any way diminish Jesus' unambiguous instructions to us.

Divine healing was part of Jesus' ministry, and it was something He expected His church to continue. Even though we may not fully understand it - to be candid, there will always be an element of mystery about it all - we must be obedient and pray for the sick.

"Very truly I tell you, whoever believes in me will do the works I have been doing, and they will do even greater things than these, because I am going to the Father." (John 14:12)

Here's the key: healing is all about Jesus and His power and authority, released to us through the Holy Spirit. It is never done out of our own strength, giftedness, or virtue. Therefore the ability to heal the sick is accessible to anyone who simply believes (even if it's just a mustard seed amount). It is all done in the name of Jesus!

Healing the sick is a calling for adults, teens, and children. It is for those who are mature in faith and those who are new to faith, and sometimes even for those who are discovering faith. Healing occurs both inside and outside of the church - in fact, it is especially common when reaching the lost. It is for major illness as well as for the common cold or achy backs, since it's the same power at work in every context.

Healing the sick should be normative behavior for every Christian.

WHAT IS THE #1 MEASURE OF SUCCESS WHEN PRAYING FOR HEALING?

When we pray for the sick, the goal is very simple: that they will experience afresh the Father's deep love for them.

Are you surprised by that statement? Perhaps you were expecting us to say that the goal is healing - after all, that's what this book is meant to be all about!

However, I can no more bring supernatural healing into someone else's life than I can deposit $10 billion into their bank account - because neither are in my gift to provide!

Likewise, you can't heal someone supernaturally. You can pray, you seek the heart of God for them, you can be attentive to the Spirit, you can learn best practices for ministry… but ultimately, the healing is not yours to give. What you can directly influence, though, is how well you reveal the love of God to that individual.

The love aspect is the piece that the Father delegates to us. It will be revealed in how we speak with the one seeking healing, how we care for them, the way in which we honor them as an individual, the tone of our voice, the tenderness with which we pray, even how we posture ourselves physically and interact if we lay on hands. We can't control the miracle, but we can bring the love.

As we become more used to praying for the sick, it becomes something that at times is more inconvenient. "Okay, I'll pray, but hurry up!" is a vibe that's easy to give off.

It sounds terrible to confess, but sometimes it's easier to rush through and 'do the task', rather than care for the individual with compassion and kindness. Yet what if the Father is less bothered about us succeeding in the task of healing, and more concerned with the revelation of His love and goodness?

As pioneering missionary Heidi Baker puts it, **"It's our job to love, it's God's job to heal."**

Of course, God is indeed good, and often He brings supernatural healing even when the one praying has no business ministering at that time due to their stinky attitude. Just because we see healing when we pray, we mustn't assume that this is a divine endorsement of what is driving us to act.

Sometimes we've observed Christians whose motivation seems to be, "I'm mad at the devil". Perhaps they've experienced loss or disappointment, or maybe they simply feel frustrated that they've not seen the answers to prayer for which they've hoped. While we should of course be opposed to the enemy, if this is our defining motive in a situation then we've now made the one we're praying for a project rather than a beloved person. Instead, we must learn how to press into healing in such a way that those to whom we minister feel authentically valued. Doing ministry out of a bruised ego is rarely a wise plan!

Instead, let's keep orientated around this prime goal: that the one to whom we minister experiences afresh the Father's love towards them.

We want people to leave our presence hopeful and at peace. Many times this will come from physical healing - again, we DO want to see lots of healings - but that piece is not in our control. However, we can operate in such a way that the individual gains a deeper awareness of the Father's presence and compassion. This means that even if they leave with the physical condition still there, they carry greater love and affirmation with them.

Healing is a manifestation of the Holy Spirit. We are the vessel, but we are not the power.

5 COMPELLING REASONS TO PURSUE HEALING

If you're wrestling with your worldview on healing the sick, here are five compelling - and very biblical - reasons to step into healing.

1. God's Character

Sickness is not from God. It is a rebellion against His character and reign, and thus He robustly and powerfully fights it.

God is a healer by name and by nature. In Exodus 15:26 God declares, *"For I am the Lord, who heals you."* We see here that one of His self-chosen names is the Hebrew phrase *Yahweh Rapha*, which translates as 'The Lord who heals you' or 'The Lord who makes you whole'. The word rapha means to restore, to heal, or to make healthy. He is the mighty God who loves to heal us - spirit, body, mind, and emotions.

The Bible repeatedly shows us that God loves to heal. Throughout the Psalms there are cries for fresh healing, as well as thanksgivings for healings that have already taken place (e.g. Psalm 6:2 and 30:2). Psalm 103:2-3 announces, *"Praise the Lord, my soul, and forget not all his benefits - who forgives all your sins and heals all your diseases."*

Isaiah and Jeremiah contain prophecies about a coming time when God's healings will occur more freely and fully (e.g. Isaiah 57:17-18, Jeremiah 30:17, and 33:6). And the classic text comes in Isaiah 53:5, which talks prophetically about the coming Messiah's saving role, which of course we see fulfilled in the work of Jesus: *"But he was pierced for our transgressions, he was crushed for our iniquities; the punishment that brought us peace was on him, and by his wounds we are healed."*

Throughout the Gospels we see Jesus revealed as the fulfiller of those prophecies. He relentlessly healed the sick, and did so proactively and with great zeal. (See, for example, Matthew 9:27-30, Mark 1:30-34, Luke 5:17, and John 4:46-54. Later we will go through His ministry more comprehensively.) And He intentionally trained and commissioned His followers to do the same thing, sharing His power and authority to drive forward this expression of His active Kingdom rule.

We can fairly conclude that sickness is not part of God's design for a good creation, that God hates disease, and that He loves to heal those who are ill.

2. Compassion

In the Gospels, Jesus was frequently moved by compassion for those fighting disease. *"When Jesus landed and saw a large crowd, he had compassion on them and healed their sick"* (Matthew 14:14).

The New Testament was originally written in Greek, and the word for compassion is the magnificent 'splagchnizomai' - try saying that loudly without spitting everywhere! It has an even better origin, as it means, "To be moved as to one's inwards, hence to be moved with compassion, to yearn with compassion".

He is still moved with compassion today, including through His Body, the church. As women and men in whom Christ dwells, it is absolutely appropriate and right to feel strong care and tenderness towards those who are sick, and to want to respond accordingly with spiritual authority and power. **We are designed to be moved with compassion when we encounter people suffering from sickness.**

3. Christlikeness

One of the functions of the Bible is to give us clear examples of how we are to live. As we read the New Testament, there is a strong and clear pathway of healing the sick: Jesus did it (e.g. Luke 13:10-13), the early church did it in Acts (e.g. 9:32-35), and ministering healing was an integral part of the life of the early church (e.g. 1 Corinthians 12:7-11).

There are about 30 different instances of healings recorded in the Gospels, and at least another 12 in the rest of the New Testament. The variety of circumstances provide numerous insights and wisdom, and the examples of Jesus and the early disciples are universally applicable. One of the ways that we are formed into the image of Christ is by being obedient to His commands and walking in His ways. As we pursue healing, not only do we minister to others, but we also become more like Jesus!

4. Connection

Healing helps people connect with God. When we are sick, our condition can easily become the dominant thought in our minds, and so encountering God in that place is hugely meaningful.

For Christians, healing encourages deeper pursuit of God and stronger commitment to serve Him. For non-Christians, it is a powerful sign of the goodness and love of God for them personally. **While they might still carry unorthodox views about God, a physical healing is something that can rapidly advance the conversation about who Jesus is and why He is worth following!** We have personally seen numerous lost friends and neighbors encounter Jesus as a result of experiencing healing prayer.

5. Command

Jesus repeatedly commands us to heal the sick. This was something that He had the 12 original disciples doing from very early in their walk with Him - notably, even from before they'd recognized Jesus as their savior and Messiah. For instance, in Mark 6:12-13 we read of the disciples, *"They went out and preached that people should repent. They drove out many demons and anointed many sick people with oil and healed them."*

Likewise, we can and should start healing the sick from our earliest days with the Lord, which is one of the reasons why we believe this is for children to do as well as adults. Healing the sick is not an optional upgrade, but rather an integral part of our walk with God. **Choosing to heal the sick is an obedience issue!**

8 COMMON FEARS AROUND HEALING THE SICK

We know now that we should pursue healing because it reflects God's *character*, it cultivates *compassion*, it is part of being *Christlike*, it deepens our and others' *connection* with God, and because Jesus *commanded* it.

Now we want to address some of the common fears around healing the sick.

Many of these concerns come from a right desire to honor God and to be people who actively let the Bible shape us. In addition, all of us are rightly alarmed when dysfunctional behavior in the Body of Christ is spotted - whether in your past or present local church, or in high-profile healing ministries.

However, it is important to keep these concerns in perspective, and not to throw out the practice of healing because some have misused or misrepresented it.

1. Fear of Error

'But is it biblical?' is the exasperated question that often comes from the back of any seminar on healing. As we have taught and trained on this topic over the decades, fear of theological error is one of the major concerns that Christians carry.

Of course, it's vital not to walk in heresy!! So throughout this book, and indeed the other books in our Naturally Supernatural series, we work hard to ground what we teach in Scripture, and also to reference others who have thought and taught with biblical wisdom on these matters.

Yet it is also important to recognize that we each carry our own bits of weak, or even heretical, theology. Whether that is the baggage of functionally valuing the Bible over the Holy Spirit, or holding too small a view of God, or overpromising what the Spirit will do in someone's life, such thinking can hold us back from stepping into healing the sick.

2. Fear of Misuse

As we travel and speak, a number of people have shared with us their bad experiences of prayer for healing. Whether it was high pressure techniques, unsubstantiated or even patently false claims of success, or anointed healers lacking any personal accountability, sadly such misuse does occur. This is never acceptable and does great damage to the cause of Christ (even if those who behaved like this were not doing so out of bad motives).

If you sense your church is ministering healing in potentially manipulative ways, we would urge you to seek godly counsel as to what to do. God might be calling you to model a different approach, or to graciously challenge those who behave this way, or even to consider leaving that church in order to go somewhere with wiser leadership. Obviously we cannot give specific direction from a book, but if there is something consistently going awry in your local congregation, then what you can't do is simply stick your head in the sand and hope it will go away.

Fortunately, in the past couple of decades there has been a steady growth of sound theology and sensible, yet faith-filled, practices, to which we hope this book will also contribute. **The best way to respond to wrong use is not no use, but right use.**

What excites us is that the same people who share stories of misuse of healing gifts are also the ones who are most hungry for biblical teaching and healthy practice.

If you have experienced misuse of healing ministry, our hearts go out to you. But please don't allow the enemy to trick you into rejecting these good gifts from God. Instead, choose the path of right use of supernatural healing, so that you can faithfully step into the call and commands of Christ over you.

3. Fear of Shiny-Suited Healers

Let's be honest: there is a cultural stereotype of what a 'healer' looks like. Shiny-suited, fast-talking, handkerchief-waving, Bible-bashing, people-pushing, cheesy-smiling, and definitely not to be trusted! These are all things that Hollywood may mock, but which also have some basis in reality.

Don't be like that!

Healing the sick is meant to be something normal and everyday, like smiling at strangers, reading your Bible, or eating lunch. Notice how many of Jesus' and the disciples' healings took place as they went about their day. Likewise, we can model a deeply passionate and faith-filled approach to healing, which is simultaneously fully accessible and above reproach to even the most hardened of cynics. This is what we will try to teach you in the practical sections of this book.

4. Fear of Pastoral Fallout

It won't be long into your journey of praying for the sick that pastoral concerns will arise. Someone very vulnerable, or worn down, or for whom the outcome really matters to you, will come across your path seeking healing. And, entirely understandably, you will immediately begin to recognize a gripping internal concern for the consequences if that healing doesn't take place.

To press into healing the sick does not mean that you have to become emotionally or relationally insensitive. There are wise ways to pray, where you demonstrate great faith and the person being prayed for feels loved and pastored through the process, whatever the outcome.

Remember the principle discussed earlier: the number one goal of prayer for healing is for that individual to experience the love of God. In other words, the ultimate goal for them is to leave knowing that they are deeply loved by the Father. That should come across in your words, actions, and the tone in which you pray, and (hopefully) their encounter with the presence of God in that moment.

Obviously you 100% want physical healing to occur! However, you can't control that outcome, but what you can influence is whether the person feels valued and loved. And that, in turn, will overcome so many of the potential pastoral pitfalls that we might fear.

5. Fear of Man (and Woman!)

Most Christians doubt that Jesus will move through them. There are all sorts of reasons for this, but the bottom line is that because we doubt God's ability to use us, we then worry a lot about what people will say if we offer to pray for them and nothing happens.

This is called 'fear of man' (meaning male and female), and it is a condition that holds back so many believers. It squashes our overt devotion to Christ ("What will my colleagues/ neighbors/ friends think of me if they discover I'm one of those 'born-again' Christians?"), our witnessing ("What if they ask me a question to which I don't know the answer?"), and, of course, our attempts at healing ("What if nothing happens?").

Fear of man will usually result in us simply not even trying, and thus our anxiety about 'failure' will override our freedom to attempt great things for God.

To be honest, the way through this stage is to fail! We need to realize that probably the outcome won't be as bad as the enemy tells us it will be, and that we must trust God for how to handle things moving forward. If we are stuck in fear of what others will say, we will never be able to step consistently into healing the sick.

6. Fear of Presumption

If you think that you're not holy enough to live a naturally supernatural lifestyle, then you're entirely correct! After all, it sounds so presumptuous to offer to heal someone from their sickness, and thus it's easy to think, 'Can that really happen through an average Jo(e) like me?'

Here's the thing: if we give in to this fear, we create a whole new level of clericalism, where only the chosen few, who are probably pastors or elders, can minister healing. This is not the biblical pattern, and is not how Jesus intends things to be in His church! (Yes, we are aware of James 5:13-15, and will share how we view this passage later in the book.)

You are not being presumptuous when offering to pray for the sick, because it is not about you. Instead, you are believing that Jesus can heal through you in the power of the Holy Spirit, and He will be the one who gains the credit and the glory.

7. Fear of Messing It Up

If you've not been part of a church culture where healing the sick was commonplace, you might be worried about 'doing it wrong'. The issue here is a lack of modeling.

In this book we are going to give you a clear five-step process that you can use as a framework to minister healing. Along the way you will also pick up lots of little tips, which hopefully will give you confidence that you can give this a decent shot! **Of course, the best coaching is experience, so it will only begin to click once you start regularly praying for the sick to be healed.** Our goal is to help you avoid as many pitfalls as possible, and enable you to create a culture in your church that is positive and helpful to this journey.

8. Fear of Change

Finally, some people simply fear the change that healing the sick represents. Perhaps you have little-to-no experience of these things for yourself - they're not really on your radar - and so you have a form of Christianity whereby you don't even think about this aspect of naturally supernatural living in your everyday life.

If that's you, own the fact that change might be tough - and then ask Jesus to help you embrace His call to heal the sick. It might be a challenge, but the fruit will be worth the cost.

DOESN'T GOD HEAL THROUGH MEDICINE?

A common objection to supernatural healing is that God uses medicine to heal the sick. And of course, modern medicine is indeed an incredible gift from God, and we should be hugely grateful for it. We know many committed Christians who are medical professionals, including one of Alex's sisters and her husband, and that is a wonderful combination. If this is your calling, keep growing in your skill set and your love for Jesus!

Like most of you, we have personally experienced the healing that comes from God's supernatural intervention through modern medicine. For Alex this was especially true when a heart arrhythmia he'd had all of his life suddenly went haywire while out on a run, and he had a near-death experience (a story for another time!). He ended up going through several treatment cycles, and eventually was fully cured through the incredible expertise of the medics at the Cleveland Clinic, the world's top heart hospital at the time. Obviously we believe that all the prayer that went on for him played a significant role (wisdom for the doctors, right treatment decisions, quick recovery, etc), but the medical procedure was a vital part of the process.

So yes, God does indeed heal through modern medicine. It's important to remember that this is not an either/or equation, but rather a both/and one! God uses *both* medicine *and* supernatural intervention to heal the sick.

OUR JOURNEYS WITH HEALING

HANNAH:

I grew up in a Christian home and we were part of a dynamic church that recognized that the gifts of the Spirit are active and available in the lives of believers today. As a teenager, my family went to an annual Christian summer camp (each family took their tent and pitched it at the county fairground - quite the smell!), where I was exposed to teaching and prayer for healing and deliverance. I was slightly overwhelmed, but also intrigued and excited by all that was happening!

I was raised with the attitude that with God healing is definitely possible, and is what we proactively pray for. In my childhood home, whenever someone was sick we automatically prayed for them. While I don't remember any miraculous healings where someone's life was changed on the spot, nevertheless it was a wonderful atmosphere of faith and obedience to have as a foundation for life.

When Alex and I married, we wanted to further build on that foundation. Not only did we minister to believers, we sought to grow in regularly praying for healing with those who didn't yet know Jesus. As we saw the healings that took place and how lives were impacted, we were adamant that this was how we were going to raise and equip our children.

It's exciting to see them now as young adults, standing on our shoulders and regularly praying for the sick, both inside and outside of the church - and more often than not seeing those people healed!

This is an ongoing learning process, and of course we still don't have it all figured out. Yet through studying the Scriptures, engaging with teaching from people further along in this journey, and much trial and error, we now have a robust theological and practical working model of what to say and do when we come across sick people. **After many years of pursuing God for healing in the lives of those around us, we have lots of wonderful stories to tell, as well as some disappointing ones.**

ALEX:

I wasn't raised in a Christian home, so the concept of supernatural healing occurring today wasn't even on my radar until after I started to follow Jesus in my mid-teens.

As I began to read the Bible, many of the amazing miracle stories from the Gospels and Acts leapt off the page, especially once I grasped that we followers of Jesus are told to do them as well! Yet when I asked older church members if someone could show me how to do miraculous works, the universal response was lots of clearing of throats, shuffling of feet, and embarrassed silences, followed by unsatisfying answers about these things not really being for us well-bred English people to do today!

The breakthrough came when I was at university, during my penultimate year there. I was powerfully healed of mononucleosis - a story that I tell in full at the start of Chapter 4.

Shortly afterwards, I spent almost four months working with African Enterprise (AE), an incredible Christian evangelistic organization that brings the Good News of Jesus to the people of Africa. My time in South Africa was while the evil of apartheid was still in force, and I was able to serve practically with teams in desperately poor townships, benefit from amazing Bible teaching and multi-racial community, and join in with various evangelistic outreaches.

One of these was with a hilarious AE Indian evangelist on a trip to an ethnically Indian area of Durban (where he loved to introduce me as "my pink-skinned friend from England!"). The power of God was clearly at work, especially when the Christians in that area invited their friends and neighbors, most of whom were from Hindu backgrounds. I watched amazed as demons manifested and were expelled with great authority, the Gospel was preached, and healings took place.

The first tangible healing I witnessed was when a young woman, whose right leg was 3 or 4 inches shorter than it should have been, had her leg grow fully out in a matter of seconds as she was prayed for. She was surrounded by her Hindu family, who all completely freaked out as they watched it happen, before coming to Christ en masse. God is so good!

That time in South Africa led to my call to church leadership (my modest plan up to that point had been to do law, enter politics, and then become Prime Minister of the United Kingdom!), and during that period Hannah and I met. In our early years of marriage and ministry, we were richly blessed to serve in several churches in England that had long histories of naturally supernatural ministry, especially the healing and prophetic gifts. We look back and see that God brought us into relationship with wise leaders from whom we learned so much.

For our part, we intentionally devoted ourselves to seeking to heal the sick as often as we could, studying the Bible, reading widely, being mentored, and going to conferences and training events. (Near the end of this book check out *Next Steps 6 - Further Reading*, with a number of recommendations on healing. These include a lovely book called *Bursting the Wineskins* by Michael Cassidy, founder of African Enterprise, where he writes about his journey into naturally supernatural ministry).

As parents, we have sought to raise our sons in these ways, so that from the very outset they have known that healing - both through medicine and prayer - is simply what we embrace. If you're intrigued about how we did that (including what worked, as well as what we would go back and change!), one of our future books in this Naturally Supernatural series will be *Raising Naturally Supernatural Kids*.

A BIBLICAL THEOLOGY OF HEALING

"Our daughter has her life back..."

I n the early days of developing a culture of prayer for healing at our former church in Ohio, we began offering opportunities to receive healing prayer during most of the weekend worship services. One Sunday a young woman, who was back home from Boston visiting her parents, came forward and shared that she had stage 4 ovarian cancer. Her doctors had told her that they had run out of options for treatment, and could now only offer pain relief and temporary fixes, rather than a cure.

Several of us gathered around and began prayer ministry, crying out to the Lord for complete healing. During that time, the young woman experienced the peace and presence of Jesus come upon her, and she left feeling refreshed and more deeply aware of His love. Over the next few days she felt increasingly energized, and at her next doctor's appointment, confusion reigned over the absence of symptoms. It turned out that the cancer had completely disappeared from her body!

The doctors assumed this must be a strange remission, so they followed her for many months, until eventually she was declared to be fully healthy. In her medical notes, the doctor recorded,

"This is a miracle, for which I have no medical explanation."

Several years later, when our family moved on from Ohio to California, we received a lovely card from this young woman's parents. They shared that their daughter remained fully healthy, was completely cancer free, and that she was marrying her fiance the following month. They concluded with the simple words: "Our daughter has her life back, and we are all so incredibly grateful to God."

HEALING IN THE GOSPELS

Vast amounts of the Gospels are given over to recording, and reflecting upon, Jesus' ministry of healing. His life was so full of the miraculous that we have only a small portion of them in writing. This reality led John to conclude, *"If every one of them were written down, I suppose that even the whole world would not have room for the books that would be written."* (John 21:25) Nevertheless, from the New Testament we learn a variety of things:

- Some stories focus on the heart and character of God, and we see His kind, compassionate, and loving nature overflowing in good deeds such as healing the sick.

- Other sections reveal Jesus actively advancing the Kingdom, as His saving power impacts people not just spiritually, but also physically, mentally, and emotionally. These sections of Scripture reveal that healing the sick is not merely an illustration of the Gospel, it is actually part of the proclamation of the Good News.

- We also receive clear teaching that directs us, Jesus' disciples across history and geography, in how we are to live today in a world that still battles sickness.

- Finally, notice that every single person who came or was brought to Jesus for help received healing. Without making this the yardstick for success for human ministry, this is still an incredible challenge to all of us who claim to represent Him in this world, whatever our experience up to now of ministering healing.

For the sake of brevity, we have simply pulled out a few of those verses here, with some short commentary. However, near the end of this book (Next Steps 5 - Scriptures to Ponder) we give you a much fuller listing of the verses by themselves, without any commentary, as a source of study and prayerful reflection.

- *"When evening came, many who were demon-possessed were brought to him, and he drove out the spirits with a word and healed all the sick. This was to fulfill what was spoken through the prophet Isaiah: 'He took up our infirmities and bore our diseases'."* (Matthew 8:16-17) Matthew clearly ties Jesus' ability to heal the sick to what He achieved on the cross, and the authority that He took back there. Thus healing is one of the benefits of Jesus' victory over sin, evil, and death (arguably sickness is just a less powerful sibling of death).

- Jesus is filled with great compassion, which motivates

Him to attack death and sickness. For instance, in Luke 7:11-15 Jesus sees a funeral for the only son of a widow in Nain. *"His heart went out to her and he said, 'Don't cry'."* Jesus touches the funeral bier and commands, *"'Young man, I say to you, get up!' The dead man sat up and began to talk, and Jesus gave him back to his mother."*

- This compassion theme repeats in Matthew 9:35-38, where we have a summary of Jesus' ministry, traveling around *"teaching in their synagogues, proclaiming the good news of the kingdom and healing every disease and sickness. When he saw the crowds, he had compassion on them, because they were harassed and helpless, like sheep without a shepherd. Then he said to his disciples, 'The harvest is plentiful but the workers are few. Ask the Lord of the harvest, therefore, to send out workers into his harvest field'."* In this text, **Jesus inextricably links the proclamation of the Gospel with healing, and directly invites us to pray for more workers who will do these partner-tasks.** Of course, once we start to intercede with sincerity, usually it will dawn upon us that we ourselves are the answer to this prayer!

- This expectation that Jesus' disciples would also heal others is woven throughout Jesus' teaching. For instance:

"As you go, proclaim this message: 'The kingdom of heaven has come near.' Heal the sick, raise the dead, cleanse those who have leprosy, drive out demons. Freely you have received; freely give." (Matthew 10:7-8)

"Calling the Twelve to him, he began to send them out two by two and gave them authority over impure spirits… They went out and

preached that people should repent. They drove out many demons and anointed many sick people with oil and healed them." (Mark 6:7, 12-13)

"When you enter a town and are welcomed, eat what is offered to you. Heal the sick who are there and tell them, 'The kingdom of God has come near to you.'" (Luke 10:8-9)

- The Gospels record a strong connection between faith and healing. For example, Matthew 9:27-30 tells us that *"two blind men followed him, calling out, 'Have mercy on us, Son of David!' When he had gone indoors, the blind men came to him, and he asked them, 'Do you believe that I am able to do this?' 'Yes, Lord,' they replied. Then he touched their eyes and said, 'According to your faith let it be done to you'; and their sight was restored."*

- When it comes to the mechanics of how we actually do the work of healing, there isn't a single model. In fact, Jesus does some pretty out-there actions!

Mark 7:33-35 records an encounter with a deaf and dumb man. *"Jesus put his fingers into the man's ears. Then he spit and touched the man's tongue. He looked up to heaven and with a deep sigh said to him, 'Ephphatha!' (which means 'Be opened!')"* Mark gives us this phrase in Aramaic, so it must have been said very memorably.

In John 9 a blind man is healed by Jesus spitting on the ground, mixing saliva and dirt into mud and putting that on the man's eyes, before telling him to wash off at the Pool of Siloam (meaning 'Sent'). The healed man subsequently found himself sent as a testimony to the religious leaders (which the rest of the chapter unfolds). Jesus' use of saliva is not quite as strange as it sounds to us today, since in that culture the saliva of a holy man was believed to be anointed. Interestingly, we have pioneer missionary friends today who have a testimony of using spit to heal a blind man in an Amazon rainforest tribe.

- There is an additional section at the end of Mark that is put in parentheses, because there is some doubt as to whether Mark wrote it or it was an insert from an early believer. In it we hear Jesus' instructions to the disciples and to those who would become disciples through them. Whether it is authentic to Mark or not, it clearly summarizes the experiences of the very early Church, and gives us great insight into healing ministry today. *"He said to them, 'Go into all the world and preach the gospel to all creation. Whoever believes and is baptized will be saved, but whoever does not believe will be condemned. And these signs will accompany those who believe: In my name they will drive out demons; they will speak in new tongues; they will pick up snakes with their hands; and when they drink deadly poison, it will not hurt them at all; they will place their hands on sick people, and they will get well'."* (Mark 16:15-18)

HEALING IN ACTS

The book of Acts records a number of healing miracles, as the Early Church followed the instruction and example of Jesus.

- When Peter heals the lame beggar in Acts 3:1-10, notice the simple command for healing - *"In the name of Jesus Christ of Nazareth, walk."* This is very much in accord with how Jesus healed - with simple, clear, bold commands. The focus is not on lengthy prayers, but on spiritual authority.

- **In the book of Acts we arguably see greater healings than Jesus did, which is what He told them - and us - to do** (see John 14:12).

Acts 5:15-16 reveals extraordinary authority on Peter. *"People brought the sick into the streets and laid them on beds and mats so that at least Peter's shadow might fall on some of them as he passed by. Crowds gathered also from the towns around Jerusalem, bringing their sick and those tormented by impure spirits, and all of them were healed."*

Acts 19:11-12 describes a similar power on Paul. *"God did extraordinary miracles through Paul, so that even handkerchiefs and aprons that had touched him were taken to the sick, and their illnesses were cured and the evil spirits left them."* Notice, though, that this was not done by Paul or the disciples for profit!

- The Early Church also recognized that the dynamic of faith was vital. In Acts 14:9-10 a man lame from birth *"listened to Paul as he was speaking. Paul looked directly at him, saw that he had faith to be healed and called out, 'Stand up on your feet!' At that, the man jumped up and began to walk."*

- At the end of Acts we see many people being healed on the Mediterranean island of Malta, and how this opened the way to the proclamation of the Kingdom. Notice how everyone is healed in this outbreak of Spirit-empowerment! We read that Governor Publius' father *"was sick in bed, suffering from fever and dysentery. Paul went in to see him and, after prayer, placed his hands on him and healed him. When this had happened, the rest of the sick on the island came and were cured."* (Acts 28:8-10)

HEALING IN THE REST OF THE NEW TESTAMENT

Themes of healing from Jesus are woven throughout the rest of the New Testament.

- The New Testament writers strongly believed that God is interested in the whole person, including our physical health. For instance:

  "May your whole spirit, soul and body be kept blameless at the coming of our Lord Jesus Christ." (1 Thessalonians 5:23)

  "Dear friend, I pray that you may enjoy good health and that all may go well with you, even as your soul is getting along well." (3 John 1:2)

- John believes that we can ask for anything in faith that is in accord with God's will (1 John 5:14-15).

- James 5:13-16 teaches about healing prayer from the elders, and we will dig into this text in the next chapter. For now, **notice that prayer for healing is normal behavior in the church, and that verse 16 (***"pray for each other so that you may be healed"***) makes clear that all believers are to minister healing.**

- Importantly in 1 Peter 2:24 we read, *"'[Jesus] himself bore our sins' in his body on the cross, so that we might die to sins and live for righteousness; 'by his wounds you have been healed'."* Echoing the commentary that Matthew offers about the healing ministry of Jesus, Peter reflects upon Isaiah 53:4, and how we are healed through Jesus' work on the cross. He presents this as a holistic healing - obviously, vitally, of our sin, but also including our bodies, minds, emotions, histories, etc. For those who would argue that healing the sick is not for today, these texts pose a significant problem. The victory Jesus won on the cross is for all time and is not something that drifted down to some lower level once the New Testament was completed. Matthew and Peter tie together forgiveness for sins and healing for our bodies to the work of Jesus in His crucifixion and resurrection. If the firstfruits of Jesus' victory are clearly available to us today when it comes to spiritual wholeness, why do we not also apply that understanding and interpretation to physical, mental, and emotional wholeness as well?

- When Paul seeks to define the value of his ministry, his insight is fascinating. He actively says it is characterized not by great teaching or deep thinking, but rather primarily by showing the power of God through healings, deliverance, prophecy, miracles, and so on.

"I will not venture to speak of anything except what Christ has accomplished through me in leading the Gentiles to obey God by what I have said and done - by the power of signs and wonders, through the power of the Spirit of God. So from Jerusalem all the way around to Illyricum, I have fully proclaimed the gospel of Christ." (Romans 15:18-19)

"My message and my preaching were not with wise and persuasive words, but with a demonstration of the Spirit's power, so that your faith might not rest on human wisdom, but on God's power." (1 Corinthians 2:4)

"For we know, brothers and sisters loved by God, that he has chosen you, because our gospel came to you not simply with words but also

with power, with the Holy Spirit and deep conviction." (1 Thessalonians 1:4-5)

"I persevered in demonstrating among you the marks of a true apostle, including signs, wonders and miracles." (2 Corinthians 12:12)

- In 1 Corinthians 12:9, Paul includes healing in the list of gifts of the Spirit *"...to another gifts of healing by that one Spirit."* Notice that if we combine this list with the one in Romans 12, we receive a beautiful mix of seemingly 'natural' and more overtly 'supernatural' activities. All of these require the Spirit's empowering if they are to help advance the Kingdom of God, and thus all of them are available to believers today. To arbitrarily pick and choose which ones are still possible today is completely without biblical foundation.

- Finally, at the very end of the Bible, in Revelation 21:4, we see a glorious glimpse of the eternity for which we all hunger and long. That will be a place where our risen Lord Jesus will *"'wipe every tear from their eyes. There will be no more death' or mourning or crying or pain, for the old order of things has passed away."* Imagine living eternally with no more sickness, pain, loss, or death! When we pray for God's Kingdom to come here on earth as it is in heaven, we are asking for this eternal and greater reality to break out around us here in this fallen and lesser reality. This includes healing the sick. Come Lord Jesus!

WHY DID JESUS HEAL?

Jesus came to definitively reveal the nature of the Father to us. In John 5:19 He said, *"Very truly I tell you, the Son can do nothing by himself; he can do only what he sees his Father doing, because whatever the Father does the Son also does."* And in John 14:9, Jesus tells Philip, *"Anyone who has seen me has seen the Father."*

If Jesus revealed the nature of God the Father, and Jesus regularly healed, then clearly healing is part of the Father's nature and will. We have a loving heavenly Father who hates sickness more than we do and who loves to see healing released into every part of His children's lives.

For Jesus, healing was a core part of His message and His ministry. When John the Baptist sends his disciples to ask Jesus if He is the Messiah, notice how Jesus wants His ministry to be assessed.

> *"At that very time Jesus cured many who had diseases, sicknesses and evil spirits, and gave sight to many who were blind. So he replied to the messengers, 'Go back and report to John what you have seen and heard: The blind receive sight, the lame walk, those who have leprosy are cleansed, the deaf hear, the dead are raised, and the good news is proclaimed to the poor'."* (Luke 7:21-22)

Healing is central to Jesus' understanding of His identity and calling. While not everything works out in our fallen world (after all, John was in prison and would shortly be unjustly executed), all disciples of Jesus should see healing the sick as central to our lives as well.

In addition, we see that Jesus healed (and drove out demons, did miracles, etc) in order to fulfill Old Testament prophecies about the Messiah. For instance, Isaiah 35:5-6 says of the coming Messiah, *"Then will the eyes of the blind be opened and the ears of the deaf unstopped. Then will the lame leap like a deer, and the mute tongue shout for joy. Water will gush forth in the wilderness and streams in the desert."* The religious leaders knew this and it concerned them that the people were reading the miracles as authentication of his credentials (e.g. John 11:45-48, Matthew 21:9-11). While demonstrating His being the Messiah was not the only reason Jesus healed, nevertheless it was part of what He was revealing.

When unpacking healing, the language that Jesus most often used was that of the breaking in of the Kingdom of God into our present day world. The concept of the Kingdom refers to God's dynamic and active rule in and through our lives, which transforms everything as the original 'good' creation intent of God is restored, often through gracious partnership with the people of God.

It's interesting that Jesus did not really teach directly about healing. Instead, He taught on the Kingdom of God, and healed everyone who came to Him.

For instance, In the Lord's Prayer this principle is seen when we are taught to pray, *"Your kingdom come, your will be done, on earth as it is in heaven."* (Matthew 6:10) The breathtaking kingly rule of God is no longer purely our future hope, because eternity is bubbling over into our lives today through the glorious life, death, resurrection, and ascension of Jesus!

Put another way, Jesus clearly believed that all the beautiful things we long for in heaven can start to be experienced today. The fullness of life He promised in John 10:10 is not just a future hope, but a present day one as well.

Jesus' very name, which means 'God saves', reveals so much about His identity and the purpose of His ministry. Since biblical thought is holistic, salvation through Jesus means that He comes to save us in every sphere of existence. He is redeeming the whole of creation - most importantly our spirits, but also our minds, emotions, and physical bodies.

At the cross we discover that salvation impacts every aspect of our humanity. This means that:

Healings are not just proof of Jesus' message, they are elements of His actual message.

Healings don't merely demonstrate that Jesus can save us, but rather they are in themselves salvation (God's redemption of everything) breaking into the here and now!

Of course, Jesus' role as our Savior is about far more than physical healing, because humanity's greatest need is for our sins to be forgiven and our relationship with God to be fully restored. Yet what Jesus offers is not some disembodied escape from the physical world. He is interested in saving the whole you: spirit, mind, emotions, and body.

BUT WHAT IF...

At this point our minds tend to jump straight to thinking about all the 'What ifs' — "What if healing doesn't take place? What if they think it's their fault? What if the family members don't see their prayers answered? What if the illness is long-term? What if they end up dying from that sickness? What if they don't have enough faith? What if they aren't healed and walk away from Jesus?"

Those are all legitimate questions, which we will address later in this book. **However, it's vital not to build our theology upon our fears and failures.** Doing so requires us to subordinate Scripture to our experiences, which is never going to produce a gospel-centered worldview. Instead, we must intentionally start with a clear-eyed understanding of God's character and actions in relation to sickness and healing.

Here is the big idea: **Jesus consistently and persistently healed the sick.** And for Him, it wasn't simply a kind thing to do, but rather it was a central aspect of His message.

JESUS' ATTITUDE TOWARD SICKNESS

Throughout the Gospels, Jesus repeatedly makes clear that sickness is an enemy of God and His people. **There is not a single incidence of Him adopting a stoical attitude towards illness.** He never viewed sickness as something that people needed to endure without complaint or a request to God for healing.

In Matthew 8:16-17, Jesus drove out the demonic spirits *"with a word"* and *"healed all the sick"*. Matthew then comments, *"This was to fulfill what was spoken through the prophet Isaiah: 'He took up our infirmities and bore our diseases'."*

Unlike some dreadful theologies today, we never catch Jesus saying something like, "Well, this sickness might be tough, but it's doing wonders for your character, so the Father says leave it in place for a little while longer." Likewise, He never responded to a request for healing by saying, "I wish I could help, but it's not yet your time for healing. The Father wants you to stick it out for a few more months."

Instead, Jesus always treated sickness as deeply and personally offensive, a violation of His good creation, and something He fiercely opposed. It is interesting that **every person who**

came to Jesus seeking to be made well experienced a complete healing — He never turned anyone down or away.

In Mark 1:41 we read, *"A man with leprosy came to him and begged him on his knees, 'If you are willing, you can make me clean'."* In the culture of the time, leprosy was viewed as something that made a person not just sick but also ritually unclean, and so lepers were excluded from the temple (the place of worship) and the village (the place of community).

Jesus' response is fascinating. He was completely indignant that anyone could for even one moment think that He was not willing to bring healing and cleansing! He didn't need to go away and ponder the case, because healing was His default answer.

"Of course I want you whole!" is the repeated message in the Gospels. In this case, He therefore touched the man and said, *"I am willing. Be clean."* Notice that by laying His hand on the leper, Jesus was not made unclean - instead, the unclean was made clean at His touch. Thus the man received not just his healing, but also a restoration into the places of worship and of community (which is why he was sent to visit the priest to have his cleansing confirmed). Such is the transformative power of Jesus!

All this leads us to conclude that **health and healing are the ordinary response of a loving Father to His children**, and Jesus went out of His way to reveal the fullness of this truth. We have a Father who loves each one of us so dearly and deeply, and thus it's never a burden or frustration to Him when we request healing, whether for others or ourselves.

SICKNESS IS NOT FROM GOD

It is surprising how many believers assert that sickness comes from God. With all the disease and illness around us, it is easy to believe this error. However, sickness entered God's perfect world through humanity's disobedience in the Garden of Eden.

While pinning the source of sickness onto God might, at first glance, seem more pastorally flexible when healing does not take place, if we dig a little deeper we'll realize that it creates more problems than it solves. By saying that God sends sickness, what sort of Father do we think God is? What kind of Savior would that make Jesus?

Following the death of his young adult daughter, theologian Ben Witherington was drawn afresh to Matthew 7:9-11. There Jesus teaches that if a human parent wouldn't give their child something bad, only something good, how much more does our Father in heaven give only good things to us? He reflected upon this in a Christianity Today article titled "God Wants to Heal Us", which carried the sub-heading, "Jesus is not the author of sickness and death — including my daughter's."

> *"That may seem obvious to some, but I was reassured in a new way that God didn't give my daughter a pulmonary embolism. He is not some selfish deity that 'needed another angel in heaven,' as one person told me at the visitation before Christy's funeral. God is working all things for the good of those who love him. But he is not the author of disease, decay, death, suffering, sin, and sorrow. He is love and the author of life - life abundant and he even promises everlasting life. His plans for us are for good and not for harm, Scripture teaches."* (Ben Witherington)

If Jesus' ministry is defined by the breaking in of the Kingdom of God, and we know that in Genesis before the Fall and in our future eternity there is no sickness, why do we ascribe sickness on earth an act of God's will? Jesus defeated death (and sickness) on the cross and through His resurrection, so why would He partner with those things on earth by sending them to individuals?

We must do better than this in our thinking and theology, and stop painting God as more akin to a Greek god who capriciously sends bad things into our lives.

Of course, we must also not swing in the other direction and pretend that all sicknesses are no longer in existence (a model that some cults take on). **Today we live in the tension of the 'now' and the 'not-yet' of the Kingdom, where sometimes we will see the sick healed, and other times we won't see that occur.**

In Chapter 5 we'll focus on how to develop both a theological and a pastoral response when healing doesn't come. But those perspectives will only be robust and meaningful when built upon a biblical understanding of God's big-picture view of sickness.

WHAT CAUSES SICKNESS?

If God does not cause sickness, then where does it come from? While there are a host of answers, it is important to have some broad categories in mind since this will impact our approach in pursuing healing.

For instance, a starting point in helping someone with a minor sports injury will likely be very different to ministering healing to a person with the physical consequences of a long-term addiction. Jesus loves to (and can) bring healing in both circumstances, but He also very much wants us to become more effective in partnering with Him, and this includes not mindlessly imposing a one-size-fits-all approach. Those who are sick are not abstract projects, but dearly-loved individuals who deserve a personal experience of the Father's loving kindness made manifest through us.

As you look at this list below of some causes of sickness, bear in mind that these are simply labels to help us gain clarity, and that real life is often more complex. For instance, you might well find several of these at work in one situation, or you could see several different illnesses colliding with several different causes, or you might not really be sure what is the cause (but that in no way stops you ministering healing!). However, it is immensely helpful to be aware of some of the range of possible causes out there.

- **Accidents** — Whether minor bumps or knocks or life-changing damage, unfortunately accidents are part of living on a fallen earth.

- **Illnesses** — In a similar way, everything from colds and tummy upsets to major conditions are part and parcel of our existence. But Jesus is definitely interested in intervening!

- **Sin** — An interesting example is the man at the pool of Bethesda. After healing him and letting the resulting media frenzy die down, Jesus finds the man later and says, *"See, you are well again. Stop sinning or something worse may happen to you."* (John 5:14) This is not a threat, but more an explanation of cause and consequence. Sin's links to sickness can be seen through bad habits that directly result in ill health (e.g. cigarette smoking that leads to lung problems). Alternatively, sin is apparent when we have worshipped the wrong things (e.g. the unbridled pursuit of success at work that leads to stomach ulcers), or when we have allowed the bondage of the enemy to come in (e.g. unforgiveness of someone that results in adverse physical ailments).

- **Spiritual Causes** — On a number of occasions Jesus identified a demon as being the root cause of a sickness. For instance, in Mark 9:25 He healed a little boy with the words, *"You deaf and mute spirit, I command you, come out of him and never enter him again."* We need to be aware that sometimes the root cause of a sickness is demonic oppression, which first needs to be commanded to leave.

- **Generational Sickness** — Sometimes there are families where the same sickness has afflicted multiple members across the generations. While not always the case, often this can be an indicator of some sort of generational curse that needs breaking before the healing will come. We'll explore this topic more in our future book on deliverance, but if you want to read more now, check out Derek Prince's excellent book, *From Curse to Blessing.*

- **Emotional Wounds** — As we go through life, it's inevitable that we'll have a number of significant, negative emotional experiences. It's common that some of these remain partially or fully unresolved. Jesus wants to heal those heart wounds, and to bring holistic healing into every aspect of our lives. Sometimes, like the tip of an iceberg, those emotional wounds break the waters of consciousness in physical illness.

A couple of years ago we were at a large conference leading a workshop on healing. After teaching on the topic for a while, we decided to publicly model how to pray for someone who is sick. In order to make it measurable, we asked for a volunteer in the audience who currently had pain in their body.

A delightful lady came forward and said that she had constant back pain — level 7 or 8 on a 1-10 scale (where 10 is excruciating). She explained that her back problems originated 14 years previously when someone rear-ended her car. Hannah prayed with her for several minutes in front of the whole room. (In those types of training situations, usually one of us will pray and minister, and the other one will 'commentate' to explain to the audience what's going on).

In this case, nothing seemed to happen! As we asked the lady for feedback (the 'Review' step in the Five-Step Model that we'll give you later in the book), it was clear her pain remained at the same level, and although her spirit was sweet and she said kind things, clearly it was not what she'd hoped for. However, as she was speaking, one of us had a sudden sense in our spirit (what Paul in 1 Corinthians 12:8 calls a 'word of knowledge') that this lady needed to forgive the driver of the car who'd hit her all those years ago.

"I'm just wondering whether you've ever forgiven the person who drove into you all those years ago?" was how the query was phrased (note the importance of not coming across in an accusatory or judgmental manner). After pausing for a few seconds, the lady said that no, she couldn't ever remember doing that. We invited her to proclaim forgiveness out loud, which she happily and easily did. Hannah then stepped into praying again for her healing, and within 15 seconds all her pain had gone, and she was stretching, twirling, and moving in ways that she hadn't done for years!

Why would it be necessary after all those years for this godly lady to need to proclaim forgiveness over the driver of a long-ago car crash in order to receive healing? To be honest, we don't know! However, we rejoice that the Holy Spirit did (and does) know, and that He was able to direct us all into the area that needed addressing in that specific instance.

This story illustrates that **the causes of sickness are varied and sometimes complex, and that the best way to unlock a situation is to be patiently attentive to what the Spirit is saying and doing.** At times you will feel a bit silly saying out loud what you sense. The logical-analytic part of your brain will be shouting, "How on earth can that be a factor?", but it is vital to learn to spot when the Holy Spirit is bringing direction.

Finally, we want to underscore the reality that **every person and every situation is unique, and when praying for the sick all of us need to be anchored in our walk with Jesus rather than a preset formula.** For instance, the lesson from this story is not that every time you pray for healing from a car crash injury, then that person first needs to forgive the other driver. Sometimes that *might* be the case — unforgiveness does seem to be one of the things that holds people back from fullness of health - but it won't *always* be true.

Sickness comes from a variety of causes, and while sometimes the relevant factor(s) will be obvious, other times they can only be discerned through the help of the Holy Spirit.

"Do not be wise in your own eyes; fear the Lord and shun evil. This will bring health to your body and nourishment to your bones." (Proverbs 3:7-8)

OBEY JESUS' STANDING ORDERS

In a disciplined yet hugely scattered organization such as the army, it is vital for troops to have clarity of direction. In the midst of the confusion and chaos of battle, the Commander-in-Chief needs to have all their soldiers operating with discipline and common purpose, so that the greater strategic plan can be achieved as effectively as possible. In order to do this, standing orders are given.

Standing orders provide individual soldiers with default commands, so that they can be operational at all times. In the absence of other specific instructions, standing orders enable the army to keep fighting the enemy, even when lines of communication back to HQ seem to be blocked.

When it comes to facing down ill health, all disciples of Jesus are likewise given standing orders from our Commander-in-Chief Jesus. As we read earlier from the Gospels, He repeatedly and unambiguously instructs us to go and heal the sick.

In fact, we can summarize our standing orders like this: unless we hear a clear direction not to heal, our permanent instruction from Jesus is to heal the sick, cast out demons, and proclaim the Kingdom.

Sometimes Christians wonder whether they can pray for healing, as if somehow we need special permission to attempt such an exceptional move, whereas Jesus makes it abundantly clear that to minister healing should be our default posture. As we go through life and walk with others as they face sickness, **almost always we are to look for the opportunity to bring the healing power of Jesus into that situation.**

So, the essence of our standing orders when facing sickness is, 'Heal the sick'!

Of course, once we understand and commit wholeheartedly to our standing orders, then Jesus can paint a little more nuance into the command.

The instruction to heal the sick does not give us license to operate in ways that are pastorally crass. Don't forget, success in prayer for healing is that the one we pray for experiences the love of the Father.

Likewise, people are not projects, and sometimes it will take a while in a relationship or situation for us first to model love and build bonds of connection, which will then create a bridge for us to offer prayer for healing. There is no standard rule for how long this will take - it could be seconds, it could be months. But we should always operate out of a heart of love and compassion, which is emotionally intelligent about the relational context.

There are also times when God seems not to bring healing. In Chapter 3 of this book we discuss some of those situations (for example, when preparing someone for death).

However, in spite of these exceptions to the rule, it is vital that we approach healing the sick with a knowledge that our standing orders from Jesus are to heal and deliver. Put another way:

> *Healing the sick should be normal and normative for a disciple of Jesus.*

If He wanted us to be cautious about dispensing healing, then wouldn't Jesus have mentioned it at some point while healing people or teaching on it? Instead, He simply tells us to heal the sick (including four out of the five times when He commissions us to declare the message of the Kingdom).

Our default setting should be to find out what is wrong, and out of a heart of love for the one who is sick, lay hands on them and invite the Holy Spirit to come with power and glorify the name of Jesus through His healing touch. We are to do that unless we sense another clear direction from the Lord.

Reflections Exercise

In the next couple of chapters we're going to move more into practical coaching on how to actually pray for the sick. But, by way of foundation, here is an exercise to help you digest some of what has been studied in this chapter.

Letting Go and Taking On

- What false beliefs about healing do you need to release back to Jesus?

- Is there any unbelief regarding healing for which you need to confess and repent?

- Where is Jesus encouraging you to more fully embrace the biblical call to healing?

PRINCIPLES FOR PRAYING

Often when we teach at a church or conference on healing the sick, we will do a live modeling of how to actually pray. This demystifies the process, which enables people to see how healing can be done in a very down-to-earth sort of way.

On one particular occasion we were in Chattanooga, Tennessee (which we still think is one of the best names for a city!), and after sharing some biblical theology on healing the sick, we asked for a volunteer who at that moment had pain in their body. A young woman came forward who had a severe headache. She shared that she had Crohn's Disease, and that she'd chosen to eat a load of gluten the previous few days. Consequently she had her worst headache ever, rating it as an 8/10 for pain.

So Hannah began praying and, as the group watched, we could 'see' the Spirit come upon this young woman and begin to minister.

One of the steps she led the young woman in doing was a simple repentance for eating too much gluten, followed by Hannah declaring the Lord's forgiveness over her. The woman had felt really guilty for choosing to eat the wrong stuff, and believed that this disqualified her from healing that day. That thinking might seem silly when written down, but it is amazing how many people come with similar thoughts in their heads, which need to be kindly dealt with from a simple biblical perspective. A scripture such as 1 John 1:9 is worth memorizing for such moments (*"If we confess our sins, he is faithful and just and will forgive us our sins and purify us from all unrighteousness."*)

After a minute or so of praying for healing, the woman said that her pain was now at a 4/10. Hannah thanked God and explained that healing sometimes comes in stages, and so she prayed again, and this time the pain went to a 0/10 and healing followed. The young woman could not believe it ("I've never seen this happen before!"), and three times she said out loud that whether this was the placebo effect or prayer, it was great. We challenged her on that misattribution of glory, and taught the group on honoring God and having faith for healing.

As a result of the woman's experience many others found their faith for healing stirred up, and the newly trained group broke up into twos and threes to pray for those who were sick.

UNDERSTANDING AUTHORITY AND POWER

In thinking about the principles of praying for the sick, it is of great importance to understand the nature of the spiritual authority and power that Jesus shares with us.

Let's start with His actual command to His disciples — which is 'heal the sick', rather than 'pray for the sick'. And there is a difference!

Jesus' instruction carries great clarity and expectation. We are not to be satisfied with intercession that sees no change.

Instead, we are told to expect to see results, to expect that the sick will be healed, as an expression of God's active kingly rule breaking into the here and now.

Of course, currently we exist in a world where the Kingdom of God is both 'now' and yet also 'not yet'. Sometimes we experience the healing, whether dramatic or gradual, but other times we are left without the thing for which we pray and long. Somehow we have to learn to live and thrive in the tension that exists between these two competing realities.

Even when we suffer great distress and disappointment, we must ensure that we build a theology from Jesus' worldview as revealed in Scripture rather than from our experiences of frustration. This is not to deny reality - we don't ignore or condemn those stories or people - yet, simultaneously, we also recognize that our personal story must submit to the grand story of Scripture and the Good News of Jesus' advancing Kingdom reign.

THE SOURCE OF AUTHORITY

We know that Jesus broke the power of sickness (along with sin and death) through His crucifixion. As we have already seen, Matthew and Peter make clear that it is by Jesus' victory on the cross that authority and power for healing are released, as there He defeated all of God's enemies, including illness.

Notice how Matthew recorded that Jesus carried complete authority to heal (e.g. Matthew 9:1-8 and 9:27-30), and that Jesus then shared with His disciples that same authority to heal the sick and cast out demons (e.g. 10:1).

This understanding of authority is unpacked early on by Matthew with the account of the faith of the Roman centurion (8:5-13), who came to Jesus seeking healing for a servant. His insight into how spiritual authority can be delegated amazed Jesus, and is thus held up as a great model for all disciples across time and history.

After the resurrection Jesus declared that *"all authority in heaven and on earth has been given to me"* (Matthew 28:18). He then declared that His authority would be shared with His followers who go and extend His active rule by making more disciple-making disciples of Jesus.

Following Jesus' ascension, this authority over sickness and the works of the evil one is tangibly released into us by the Holy Spirit. **As Spirit-filled followers of Jesus, we can now exercise the gifts of the Spirit (such as healing) in ways that align with the fruit of the Spirit (we minister healing with love, kindness, patience, and so on).** The authority to heal has already been given to us by Jesus, which means that our task is simply to exercise that authority on His behalf!

We heard it described this way: Imagine that someone decides to give you a car, which arrives on your doorstep brand new and fully paid for, all ready to go. Wow, what a gift! And here's the thing: your next step is not to go down to the car dealership and apply for a loan! The car has already been paid for.

Jesus has already bought healing through what He did on the cross. Unfortunately, though, to go back to the car analogy, much of the church seems to have mislaid the keys! Yet even though we may not yet know how to access the fullness of that gift, Jesus has already paid the price.

We seek to bring healing to the sick, therefore, because we want Jesus to receive what He has already paid for. It is all for His glory. While in this world not everyone will be healed, nevertheless some will be. As with all gifts of the Spirit, we can grow both personal and church-wide competency in ministering healing.

So know this: **if you are a disciple of Jesus, you carry incredible delegated authority and power to heal the sick!**

AUTHORITY REQUIRES SUBMISSION

To exercise authority in any realm requires that the one doing the task is submitted to the original source of that authority. For instance, a town mayor only has authority insofar as it has been delegated to them by the people of that community, who rightly expect such authority is to be exercised on their behalf and for their benefit.

Often we carry an incorrect understanding of the gifts of healing. We think it's something that only a very select group of people can do, and either someone 'has the gift' or they don't. This is basically gunslinger theology: a slightly oddball or mystically anointed person expertly wields the weapon of healing, and the rest of us take cover and hope it all works out!

A healthier and more biblical view is to understand that any believer can exercise the gifts of healing, because it is not about their personal authority or worthiness. Instead, if we walk with Jesus, we can be empowered by Him to do these things on His behalf and for His glory. While we can develop best practices and so on, ultimately the empowering comes from Jesus.

Our part is more about lowering our internal resistance to the flow of the Spirit's power, so that God can do more through us. But this is only possible through a heart that is increasingly submitted to Jesus.

To grow in authority, we must go deeper into submission.

Submission also means that we recognize that God's authority is greater than the enemy's, even when expressed in a terminal illness. When we encounter someone who is sick, the temptation is to start from a place of fear, sentimentalism, anger, or even ego. Instead, we must begin from a place of focusing on Jesus and who He is.

This means we refuse to give reverence to the sickness or be overly impressed by it ("Oh my, you've got cancer"), and instead we start with who God is, and what He tells us to say and do. The submission piece is that we choose to honor Jesus far more than we honor the actual illness. We must consistently be more impressed by who Jesus is and His nature, than by the sickness or the work of the evil one.

Authority also increases when we are promptly obedient to the Father. This applies in the broader sense of us actively seeking to be obedient to the general guidelines given by Scripture, such as the Ten Commandments. In addition, it is especially true in our responses to the specific, day-to-day promptings of the Holy Spirit. For instance, when we sense Jesus asking us to give money to one person, or speak words of encouragement over another, our simple acts of obedience steadily help grow our spiritual authority.

To increase our authority, we should seek to respond to the Lord with obedience. **Remember that obedience is God's love language!**

DO I NEED THE GIFT OF HEALING?

Sometimes believers can become stuck on whether or not they have the gift of healing, seeing that as an essential gateway into ministering healing to the sick. Put negatively, some say they won't even attempt to heal because they don't think they have the gift of healing. We would politely suggest that this is not Biblical thinking!

As mentioned already, there is a spiritual gift of healing (see 1 Corinthians 12:9), and some Christians will have that gift. This will be demonstrated by an individual seeing more healings taking place than most of their peers do. If this becomes your story, keep humble and keep ministering for the glory of Jesus.

However, any of us can be used by God to heal the sick person in front of us, even if we think we don't have the gift of healing. **It's not a question of whether or not you think you're gifted, but whether or not you're willing to be obedient to Jesus' crystal clear instruction: heal the sick.** As we step out in faith, we trust that He will empower us in that moment.

Most spiritual gifts work this way. Some people have the gift of evangelism, and consequently will see more people come to faith in Jesus than is usual. However, all of us are explicitly commanded to be Christ's witnesses to the ends of the world, as we do the work of an evangelist. Other people have the gift of hospitality (1 Peter 4), and will do an incredible job of welcoming and loving their guests. Yet all of us are instructed to be hospitable and to open up our homes and lives to others.

Don't fall for the temptation of obsessing about exactly which spiritual gifts God has given to us personally, which leads people to sit around waiting for the heavenly zap before they do anything! Instead, recognize that we are all called to minister broadly, and often it is only as we go and put ourselves in situations that stretch our faith that the Spirit will supernaturally empower us. If over time we spot repeated patterns of ministry, we might conclude that we have certain spiritual gifts. However, this knowledge is not necessary to obey Jesus' call to serve and bless others, and specifically to heal the sick.

COMMAND VS. REQUEST PRAYERS

Our true beliefs about spiritual authority are often revealed by the way in which we actually pray for healing.

Specifically, there is usually a greater impact when we pray clear command prayers for healing. This means something like, "We command this sickness to fully leave Mel's body".

However, if you listen in to many prayers for healing, what they actually turn out to be is hesitant request prayers, where people petition and ask God if He'd perhaps like to heal as a special favor.

Obviously we must always come to the throne of God with great humility and respect. Simultaneously, though, our honoring of Jesus means that we must be obedient to exercise the authority that He has *already* shared with us as believers, and to stop acting (and praying) like that hasn't yet occurred. Jesus has delegated His authority to us for a reason: to help us grow and step into the ministry that He desires us to exercise today, so that we can more effectively advance God's Kingdom rule now.

When we command healing, we are speaking to the sickness and demonic realm behind it, breaking its power and declaring the life of God. At first blush this sounds like crazy presumption on our part! Yet the whole point is that we are not doing this in our own strength, but as stewards and ambassadors of the authority and power of Christ.

From Genesis 1 onward the spoken word has great power, and if we are walking in step with the Spirit in doing the work of the Kingdom then this should be our model today.

As disciples of Jesus we straddle both the natural and the supernatural worlds. One way we bridge them is through the words we speak out of our authority as those who are already seated with Christ in the heavenly places, so that we might display the kindness of God, in this case, to faltering bodies (Ephesians 2:6-7).

If you still struggle with this concept, consider some examples of how Jesus uses commands to bring healing.

- When Peter's mother-in-law had a high fever, Jesus *"bent over her and rebuked the fever, and it left her."* (Luke 4:39)
- Jesus was asked by a leper if He was willing to heal him. *"Jesus was indignant. He reached out his hand and*

touched the man. 'I am willing,' he said. 'Be clean!'" (Mark 1.41)

- When a paralyzed man is laid before Him, Jesus commands, *"I tell you, get up, take your mat and go home."* (Mark 2:11)
- When He encounters a deaf and dumb man in the Decapolis, we read, *"He looked up to heaven and with a deep sigh said to him, 'Ephphatha!' (which means 'Be opened!')."* (Mark 7:33-35)
- In Luke 7:1-10 the Centurion is commended for his faith, because he asks simply for Jesus to *"say the word and my servant will be healed."* That is the attitude that Jesus wants us to imitate: a robust faith that succinctly and boldly commands healing.

In case you feel that only Jesus can operate this way, here are some examples from Acts.

- Peter heals a lame beggar with the words, *"In the name of Jesus Christ of Nazareth, walk."* (3:6)
- Peter heals a man in Lydda who had been paralyzed for 8 years. *"'Aeneas,' Peter said to him, 'Jesus Christ heals you. Get up and roll up your mat.' Immediately Aeneas got up."* (9:32-35)
- Paul met a man who had been lame from birth. *"He listened to Paul as he was speaking. Paul looked directly at him, saw that he had faith to be healed and called out, 'Stand up on your feet!' At that, the man jumped up and began to walk."* (14:9-10)
- Paul recounts his conversion story, including when Ananias healed him: *"He stood beside me and said, 'Brother Saul, receive your sight!' And at that very moment I was able to see him."* (22:13)

If we are taking the Bible seriously as our handbook for life, then we must move away from casual, half-hearted statements. **We have the delegated authority to proclaim with faith the now-word of God, to say what He is saying in a specific situation, such as, "Cancer, be gone!"** To do this, we must listen to what Jesus is saying, and declare it boldly.

Remember, this is a power confrontation, since disease represents a blatant opposition to the Kingdom of God (because there is no sickness in heaven). Our task as faithful disciples is to confront sickness as something that needs to retreat before the advancing reign of Jesus.

However, this does not mean that you have to shout, snarl, or stare crazily! Remember that Jesus often stepped away from the crowd when healing an individual, so as to not embarrass them or demean their value. Likewise, we speak crisply to the sickness, yet in a manner that demonstrates loving kindness and respect to the person seeking healing.

HOW TO COMMAND

While we see commanding healing as our default option (in the absence of God giving any other instruction), there are times when it feels especially useful.

These include when we are called upon to:

- Break a curse or vow
- Cast out an afflicting spirit
- Pursue a healing that seems to be held up by spiritual warfare
- Offer healing in an evangelistic situation

Some practical phrases that we might use include, "I break the power of this condition in the name of Jesus", or "Be healed", or "All sickness is to leave now", or "We break the power of any afflicting spirit, and command you to leave in Jesus' name, and speak total healing into this body", or "We speak complete healing to every joint, muscle, or ligament that is in any way out of place."

With cancer we have found that it can be more effective to also curse it. We might say something like, "We curse this cancer, and command it to shrivel up, shrink back, die, and to fully leave this body".

Please don't take these words as a new legalism, or a set of must-say phrases! While some are clearly echoing actual words of Scripture, others are more the result of trial and error over the years in actual situations we've faced. **Experiment with words that feel authentic to you and your context as you pray for sick people**, and you'll find that certain turns of phrase help stir faith in others (and yourself), and produce a greater impact in terms of healing.

SO IS INTERCESSION WRONG?

NO! Absolutely not!!

We must intercede for those who are battling illness. For instance, a friend who is recovering from surgery needs prayer for their body to fight off infection and heal up quickly. A neighbor whose child has the flu needs intercession for the sickness to be gone, for no-one else to become infected, and for wisdom in helping the child drink enough fluids and return to eating. We can pray for those we know who are headed into hospital for a procedure, for safety, success, favor on the medics, and a speedy recovery. All of those are good and appropriate things to do.

A posture that we will sometimes adopt is one of prayerfully lifting that person before the throne of Jesus. We imagine Jesus with the person - what is He saying, what is He doing?

We then pray in accordance with how we sense the Lord is leading, and what He is doing.

However, many Christians stop at intercession, and that is the mistake. When asked to pray for healing, Jesus expects us to exercise the spiritual authority that He died to release to us. Too many Christians are so fearful of failure (and the subsequent disappointment or tough questions) that they prefer to disobey Jesus and not step into that authority and power.

SHOULD I ALWAYS INCLUDE 'IN THE NAME OF JESUS'?

Obviously every naturally supernatural act that any of us do is through Jesus. Reflecting on this truth, sometimes people will ask, "Should I always include 'In the name of Jesus' when I pray for healing?"

While it is always wonderful to have the name of Jesus on our lips, there is not a legal requirement to justify all our prayers by tagging on 'in the name of Jesus'. As long as we are submitted to Him and it is His authority that we are using rather than our own, then a simple command such as "Back, be healed" should suffice.

You will see in the example phrases in this chapter that some include the name of Jesus and some don't. These were put together randomly, so don't go building some theology out of which types of prayers overtly call upon His name! **Instead, the principle is that if our hearts are devoted to Jesus, if we seek to hear and obey Him in every situation, and if we operate within an accountable Christian community, then probably we won't go too far wrong in the exact details of what we pray in each specific moment.**

One nuance to that principle is that we will try to say the name of Jesus at least once when praying for healing for (or alongside) a lost person. This is simply to make clear who the prayers are being prayed to, and whose power is at work, so that when the healing comes they know that it wasn't just some general vague prayer to the universe!

WHY WE BAN 'IF IT BE YOUR WILL' PRAYERS

In the Lord's Prayer (Matthew 6:10) we are taught to pray, *"Your kingdom come, your will be done, on earth as it is in heaven."* Our vision for the whole of life needs to be in Long Beach/ Oklahoma City/ London/ Timbuktu or wherever we live, as it is in heaven.

The 'will' part of this phrase is defined not by doubt ('*if* it be Your will') but by what we know of His eternal Kingly desires. Therefore, if something is a facet of the Kingdom, then it is God's will. For instance, we do not have hesitancy in praying for a person's salvation, since we know from Scripture that God *"wants all people to be saved and to come to a knowledge of the truth."* (1 Timothy 2:4)

If in heaven there is no ill health or death, then we should fight against sickness on earth, because the Kingdom coming in greater fullness today *is* God's will. Of course, we are still caught between the now and the not yet of the Kingdom, so it won't always happen how we wish. Nevertheless, our starting posture needs to be grounded in the possibilities of Kingdom life, so that we walk by faith and not by sight.

Notice that no one in the Bible ever prayed 'if it be Your will' with regards to healing (or anything else for that matter). Instead, they prayed with great boldness. (Occasionally someone initially petitions God to grant him or her power, but always the healing itself comes from a spoken command, or an intermediary action that represents such bold faith.)

Generally we find that 'if it be Your will' prayers are one of the quickest ways to dampen faith, since they sow a hefty dose of doubt into the situation. Instead, let's do things biblically, and stop using this preemptive excuse to try to cover our backs in case healing doesn't happen!

WHEN SHOULD WE NOT OFFER PRAYER FOR HEALING?

Our standing orders from Jesus are that we are to heal the sick. Jesus makes this clear on repeated occasions (as shown in Chapter 2), and it has been the experience of Christians from the early church onwards that God loves to heal through our prayers.

So, we should always generously offer to minister healing... unless Jesus says otherwise. That last phrase might sound surprising, but there are some occasions when to pray for healing would be inappropriate.

There is no definitive list, since every situation is unique. We must always focus on listening to Jesus each time, rather than relying upon rules or even our previous experience.

Nevertheless, to help be better prepared, here are four situations where we've found that *sometimes* the Lord seems to direct us not to pray for healing.

1. When Death Is Close

There is time to prepare someone for death. None of us will live forever in this world, and part of the pastoral responsibility of the church is to play our part in readying someone for that great transition. Simultaneously, our actions at those times are also of huge importance to the family and friends of the loved one.

With a longer-term terminal illness or an extended stay in intensive care, sometimes the realization can come that a person is clearly going to die. In those profound moments - where time seems to stand still, and the gap between earth and eternity seems so thin - to insist on carrying on praying for healing will at best be tin-eared, and at worst be deeply unhelpful to the person or their loved ones.

Obviously that is a very sensitive judgment call, but our top motive in this ministry is to help others experience afresh the Father's love. Our focus might rightly shift to comforting, guiding, and praying with the one close to death, reassuring them of the love of God, and the truths of the Gospel message. Likewise, we may be able to minister to and pray with family and friends in unexpectedly intimate ways.

If you find yourself in such a situation, keep very attentive to the Lord, and be extra sensitive to those around you and what they seem to be saying and indicating. Do not blindly plow on with insisting on healing when, however tragically, a different type of ministry of the Spirit is required.

2. Hostile Atheists

If you're with a non-Christian who is extremely hostile to the Gospel, an offer to pray might simply bring a rebuke and possibly hinder their openness to Christ. Some people are just not in a place to receive healing. Remember, Jesus (and the disciples) only healed those who came to them and asked, or in whose lives the Spirit was already at work.

However, it's worth noting that such a response would be unusual. We have found even those who seem far from God to be very open to healing, and have seen many healings take place amongst those who don't currently follow Jesus.

3. Perception of Sickness

There are some conditions that from the outside might look like they require healing, but from the perspective of the individual and their family, they really don't.

For example, we have had a number of conversations with parents of children who have a milder form of autism, and they see so many beautiful aspects of their child's character that might be removed if that condition was not present. To simply march in and try to heal what doesn't need (or seem to need) healing violates so many important values.

We must always honor the individual, operate as servants, start with listening, and overflow with genuine love.

Christians who fail to do these things can cause a lot of hurt and anger, as they dishonor the rich tapestry of life that God is threading together in that unique person.

Does this mean that we should reflexively refuse to pray for someone with (for example) autism who seeks healing? The problem is that we're talking about shades of grey here - life is not always as binary as 'healthy vs not-healthy'. Someone with autism might pursue healing for some or all aspects of that condition, and in all likelihood they would be able to express what it is they're seeking. Our task is to respond out of love and servanthood to that request, without insisting that everyone else with that condition follow the same pathway.

We need to ask the individual or their parents (as Jesus often did), 'What would you like God to do for you?' This is where we go back to the foundational principle of ministering healing - what would be the most loving thing to do? How can we best represent the Father's love for this precious individual? How do we play our part in helping them become more aware of their intrinsic value to God? It is from that place that we must operate.

4. Prayer Exhaustion

Finally, there are situations that require an increased sensitivity when considering prayer for healing. For instance, there are people who have had a long-term illness or condition, and they have been prayed for many times. Sometimes their hesitancy about future prayer is that they have been treated in insensitive ways by Christians, or they have been disappointed by lack of healing one too many times.

When we spot someone in a wheelchair and want to pray for them, let's first pause and ensure that we are walking in step with what the Father is doing. There can be a danger of us doing a version of Christian virtue signaling by publicly seeking to heal those with the most prominent needs. Sometimes that might be the Spirit's leading, and sometimes it's simply our ego wanting another notch on our spiritual belt (and that's not the belt of truth, by the way!). Or, to put it in a simpler way, maybe that desire is from your own heart rather than what God is doing or what the person is ready for.

So while in most of this book we want to encourage greater faith and boldness (let's face it, most churches need more of both of those things!), consider this the cautionary tale. **In pursuing healing we are not to swing the pendulum into scalp-hunting territory, running full steam ahead in our own strength rather than listening for what Jesus is saying.**

People are not projects. Your task is not to minister healing to the maximum number of people as possible, operating with a production line mentality. Quick insight: you are not that important! The Kingdom will still advance without your help! Instead, your task is to represent the love of the Father for as long as is appropriate to the situation at hand. Sometimes that might be a quick touch, other times it will be a five or ten minute interaction, and occasionally it will be a very lengthy and extensive sharing and caring.

Remember: it is all about love - love of God, and God's love for the individual. It's not about you!

FAITH AND HEALING

A quick pop-quiz: In the Gospels, can you name the two occasions when Jesus was amazed?

Extending that question further, what would it take for us to amaze Jesus? As we will see below, both of these times were to do with the people's faith (either the abundance or lack of it).

It is clear throughout the healing ministry of Jesus that faith was one of the key elements in play. As New Testament scholar Ben Witherington comments, "There is a positive correlation between faith and healing."

For the sake of clarity, we define faith as, "To be firmly persuaded of God's power and promises to accomplish His will and purpose, and to display such a confidence in Him and His Word that circumstances and obstacles do not shake that conviction."

There are times when the Scriptures seem to suggest that a lack of faith can prevent or limit the healing power of God. For instance, in Mark 6:1-6 we read of Jesus being rejected by the people of His hometown Nazareth, and that,

> *"He could not do any miracles there, except lay his hands on a few sick people and heal them. He was **amazed** at their lack of faith."*

A clear connection is drawn between a lack of faith and a lack of miracles, including very few healings taking place.

The account of the Centurion asking Jesus to heal a beloved servant presents a positive and aspirational model of faith. We read in Matthew 8,

> *"The centurion replied, 'Lord, I do not deserve to have you come under my roof. But just say the word, and my servant will be healed. For I myself am a man under authority, with soldiers under me. I tell this one, 'Go,' and he goes; and that one, 'Come,' and he comes. I say to my servant, 'Do this,' and he does it.' When Jesus heard this, he was **amazed** and said to those following him, 'Truly I tell you, I have not found anyone in Israel with such great faith.'"*

The Centurion is held up as an example of faith because he clearly sees that Jesus has all power and authority, which He can release at will through a simple declaration. There is a unique combination of great humility and total trust at work here. The man's military training would have helped, since complete obedience to and confidence in your leaders are fundamental to the chain of command working successfully in a battle.

The way to amaze Jesus is by our faith for healing. Let's aim for the positive kind of amazement!

WHOSE FAITH?

"Faith is the medium through which God releases his healing power." (John Wimber)

The importance of faith raises the question of who needs to be exercising it. We see four main options, although it should be noted that often faith will come from a combination of several, or even all, of them.

1. The Person Seeking Healing

Obviously if the one seeking healing has strong faith, that is going to be a good thing! A couple of examples help our thinking:

- Three of the Gospels record the account of blind Bartimaeus shouting loudly and persistently for his sight to be restored (Mark 10:46-52). In fact, he is so noisy and insistent - *"Jesus, Son of David, have mercy on me!"* - ignoring the shushing of those nearby, that his crying out halts the whole parade. You could call it the shout that stopped God! Bartimaeus demonstrates both faith and persistence (the two key elements to which Jesus often returns), and he is healed with the words, *"'Go,' said Jesus, 'your faith has healed you.'"*

- In Mark 5:25-34 we 'hear' the thoughts of the woman who had suffered from a dozen years of severe bleeding, when she determines, *"If I just touch his clothes, I will be healed."* And gloriously, the moment she does just that we discover that, *"Immediately her bleeding stopped and she felt in her body that she was freed from her suffering."* Yet this bleeding had not only brought great physical pain, but also emotional and relational heartache, since her community viewed a bleeding woman as ritually impure, leading to her exclusion from both social gatherings and also worship at the Temple. After He is touched in faith, Jesus is aware that power has gone out from Him. (As an aside, some people who are experienced in ministering healing to others have a similar sensation.) Jesus then makes a big scene in order to have the woman identify herself publicly. At first glance this seems unkind, until we listen in to Jesus' words where He publicly pronounces her clean and thus redeems her social and worshipping world, bringing emotional healing alongside her physical wholeness.

2. Friends and Relatives of the Person

There are numerous examples of the faith of those accompanying the sick one being the thing that is critical.

- The classic example is the four friends of the paralyzed man, who are so determined in their faith that they break a hole in the roof in order to lower him in front of Jesus. *"When Jesus saw their faith, he said to the man, 'Take heart, son; your sins are forgiven.'"* (Matthew 9:2) It is the persevering faith of the friends that invites forgiveness, and then healing, from the heart of the Lord.

- There are several accounts of people asking Jesus to heal someone who was not physically present. Jairus the synagogue leader (Luke 8:41-42, 49-56) models this, risking his status in society in order to publicly come to Jesus. Even when a message arrives that the little girl has died, Jesus breathes life into that faith. *"Hearing this, Jesus said to Jairus, 'Don't be afraid; just believe, and she will be healed.'"*

3. Those Who Are Praying for Healing

We who pray for the sick must come with eyes of faith. It is so tempting to be controlled by what is going on in the natural world, whereas we are to be the ones who *"walk by faith, not by sight"* (2 Corinthians 5:7).

- Visiting the town of Nain in Luke 7:11-15, Jesus encounters a funeral for the only son of a widow. *"His heart went out to her and he said, 'Don't cry.'"* Jesus then touches the dead body (interestingly, a good Jewish rabbi would never touch anything dead to prevent spiritual defilement - but Jesus is the one who makes everything holy by His touch!), and commands, *"Young man, I say to you, get up!"*. This is a beautiful example of the faith of the one praying, following the lead of the Spirit who reveals what the Father is wanting to do in a particular situation (since presumably Jesus saw many funerals during His life and didn't interrupt them all).

- When Paul was shipwrecked on Malta, Publius, the chief official of the island, hosted everyone for three days. In Acts 28:8 we read, *"His father was sick in bed, suffering from fever and dysentery. Paul went in to see him and, after prayer, placed his hands on him and healed him."* The implication is that Paul wanted to proactively bless Publius, and when he hears of the sick father, his faith is stirred for healing.

4. The Church Community

There is a great responsibility on local groups of believers to be intentional about growing in persevering faith and humble expectancy. It is often easier to choose doubt, cynicism, and reluctance, but such things are death to a culture of faith.

- As we discussed above, the synagogue at Nazareth took great offense at Jesus' teaching and miraculous powers, because they knew Him solely as a local lad made good. Matthew sadly comments in 13:58, *"And he did not do many miracles there because of their lack of faith."* A community of faith that lacks faith is of little use to anyone. As the church, we are to become an environment where the water table of faith is constantly rising, so that even the most worn-down and leaky vessels are lifted along with the whole group.

- In the Gospels, Jesus always sent the disciples out in groups to heal the sick, drive out demons, and proclaim the presence of His Kingdom. In other words, we are meant to do this in the context of community. When we're alone it's easier for the enemy to whittle down our faith. The Christian life is not meant to be done solo, and one of the key roles of the local church is to be a hothouse of the faith that perseveres.

- James 5:14-15 reads, *"Is anyone among you sick? Let them call the elders of the church to pray over them and anoint them with oil in the name of the Lord. And the prayer offered in faith will make the sick person well; the Lord will raise them up. If they have sinned, they will be forgiven."* The picture here is of submission to the authority of your local church. The type of healing prayer described here seems to be an official act of the church, rather than the individual who prays. The one who is sick has to go in humility and say, 'I need you, as representatives of my church family, to pray for my healing.' However, this teaching in no way restricts healing authority to the elders. In fact, the very next verse (16) says, *"Therefore confess your sins to each other and pray for each other so that you may be healed. The prayer of a righteous person is powerful and effective."* This faith of the local church community, that is modeled and stirred up by the elders, is then released to all those who look to those elders for leadership. All of us can be the righteous ones who minister effective prayers of healing.

HOW TO PRAY WHEN THE RECIPIENT DOESN'T HAVE MUCH FAITH

Sometimes you will be asked to pray for someone who clearly doesn't have much (if any) faith for healing. Likewise, those accompanying them might be full of doubt or despair. Alternatively, we can be looked to for help, when on the inside we're thinking, 'Gosh, I've never seen God do anything like this before!'

Here are a few things to bring to mind:

- God is the one who does the supernatural activity, so we should always go back to His character and nature. What sort of view of God do we have? Do we think Jesus begrudgingly doles out the minimum number of healings He can get away with, or is His heart one of love, generosity, and goodness?
- Faith is having eyes to see into the invisible spiritual realm, recognizing it as being the greater reality (see Hebrews 11 for more on this). When we encounter sickness and pray for healing, we are leaning into that overarching rule of God. Therefore prayer for healing is inherently an act of faith, and should not be governed by what seems possible according to the regular rules of the material world.
- Faith is something that can be caught from others (likewise, so can things like fear or doubt). When you are with someone who is low on faith, that is your opportunity to change the prevailing climate around them. Be a spiritual thermostat, not a thermometer!
- If you're a Christian, then Christ lives in you, and His faith is always perfect. When He was on earth He always trusted the Father, and now in heaven Jesus remains the faith-full One, who loves to pour His faith into us.

Some practical steps you can take include:

- Remember that someone who has had their faith ground into the dust over a long illness needs to be loved and not rebuked. As you come alongside, internally ask Jesus to give you His heart of compassion and love for them, so you might shepherd them with both tenderness and faith. Jesus is the one who won't break a bruised reed, or snuff out a smoldering wick, and neither should we.
- The fact that they are there means that someone, somewhere, has some faith! If you can spot where that is, be sure to affirm it out loud.
- Allow the person to be candid about holding in tension their faith and their frustrations about not being healed up to that point. While those disappointments don't define our theology, you make it a safe place for them to let their guard down and encounter Jesus if you are comfortable with some raw honesty.
- Having said that, it's rarely helpful for the one seeking healing to go into a long recitation of the difficulties of their medical journey. This is not because we don't care, but because it only feeds doubt and despair. Beware accompanying friends and family who've taken up offense on behalf of the one who is sick. Loving compassion means that we don't feed this negativity, but instead we help them catch a fresh glimpse of the greater truths that are so easy to lose sight of when in pain.
- Remove friends and family who are too caught up in doubt and disbelief. Jesus did this on a number of occasions (e.g. Mark 5:35-43). We might say, "We're just going to step over here by ourselves with Nicky to pray, we'll just be a couple of minutes." Obviously,

do this kindly, but only having those with faith around you does help the one for whom you are praying.

- Start by declaring truths about who God is, His attitude towards sickness, and examples of Him healing (both biblical and from your own experience). Praise Him for who He is, and thank Him for what He has already done in that person's life and in the testimonies you've recounted. If you have the gift of Tongues, it can be helpful to quietly pray that way as well.

- Ask Jesus to release prophetic words or insight to help unlock the situation. These might speak to the causes of the lack of faith (and thus release healing into those memories), or inspire the person today, or simply guide how you pray. We speak a lot more to this in the next chapter.

- Record what you pray onto a Voice Memo on your phone. This makes what goes on so tangible, as the recipient can listen to it again (sometimes they'll do so multiple times), which will leave them feeling loved and comforted, and with their faith boosted.

Finally, if you personally need stronger faith to generally minister healing, here are some strengthening exercises:

- A study of Jesus' healing ministry from the Gospels.
- A study of healing in other places in the Bible.
- Read stories of healing from contemporary sources, or from church history - but don't just stick to your own country, culture, or church tradition! (There are some suggestions in the *Further Reading* section of this book.)
- Hang out with people who operate more effectively

in this ministry. See if you can join with them when
they are ministering healing.

- Go to places where healing is happening, and allow
your faith to be built up.
- Remind yourself that cynicism is not a fruit of the
Holy Spirit!
- Ask Jesus to grow your faith for the naturally
supernatural to occur through you. This might
include repenting of beliefs and theologies that have
built doubt rather than faith.
- You were designed for answers to prayer (hence we
are repeatedly invited by God to be prayerful people -
He does not intend for us to live constantly
frustrated!) However, it's helpful to have a variety of
different prayers at any given time 'on the table'
before God. Some of these will be short-term / quick
wins (like a headache to be healed), which give you
ongoing fresh answers, which will do wonders for
your faith. Some other requests might be long-term or
major requests, and so it is important to keep our
faith and hope high for those situations from other
answers to prayer. Therefore, make sure you are
praying for smaller or short-term things as well as
long-term or major issues.

Faith is a huge topic - after all, it is both a gift and a fruit of
the Holy Spirit. We will write a lot more about it in our book
in this Naturally Supernatural series, *Living a Miraculous Life
of Faith*.

A PRACTICAL GUIDE FOR HEALING THE SICK

When I (Alex) was at college, I had a nasty run-in with mononucleosis. The illness came out of the blue, wiped me out for several months, and was followed by a frustrating season of dropping in and out of extreme tiredness and lack of energy.

As I sat in a church service one Sunday evening, I was struck with the realization that I should ask my pastor, Roger, if he 'did' prayer for healing. He of course said that he'd love to pray for me and arranged a time during the week to visit with him.

I was so excited! Obviously I was hoping for healing, or at least some improvement, but more than that, I eagerly awaited my induction into The Special Prayer For Healing™, which I presumed had been carefully passed down by word of mouth, person to person, through generations of Christian leaders. I wondered if it might even date back to the original 12 disciples!

When the time came, Roger asked me about the illness, and what my journey had been. By the time he said that we would now pray, I was on tenterhooks.

Yet what followed was such a surprise. He laid hands on me, prayed a brief-ish prayer that was clearly extempore, and then sat back. "Is that it?" I thought, just about demonstrating enough self-control not to blurt that out loud. Yet I did feel an unusual sense of God's presence, and I knew Roger to be a man of God, so I graciously decided to give him the benefit of the doubt, for a few days at least.

And then, over the days that followed, my health rapidly improved. My energy shot up, my appetite suddenly returned, and within a couple of weeks every sign of the illness disappeared. I knew that Jesus had healed me - and I realized that the 'formula' was far more to do with a personal connection with the Holy Spirit, rather than saying the prescribed right words.

"HOW DO YOU ACTUALLY DO A HEALING?"

Recently, after we'd taught a short training session on naturally supernatural living, the question on the lips of a rather frustrated lady at the coffee break was, "But just how do you actually *do* a healing?" She clearly felt that we'd focused too much on the theory, when all she wanted was a model that she could take and implement, and a simple set of words that would make all the difference.

Those are not wrong things to desire, and there is best practice from which we can learn. After all, if our goal is that the person for whom we are praying experiences the Father's love for them, then there will be more and less wise ways to help facilitate that encounter.

Therefore, in this chapter we are going to share some very practical suggestions, including a five step process that provides a hugely helpful framework.

However, a health and safety warning comes first!

There is no secret, sacred, or magic formula for healing the sick.

The simplest version is for you to put a hand on the person's shoulder and declare, "In the name of Jesus, be healed!"

Sometimes that will be all that you need to do. For a Christian, the essence of healing the sick really should be that uncomplicated.

As the one praying, always remember to ask Jesus for specific direction, recognizing that He will guide you each time. Keep in mind the multitude of ways and approaches that Jesus took to healing the sick (spit and mud, anyone?!), and that will remind us not to put our trust in formula over faith, or rules over reliance upon the Spirit.

In many situations you will sense that the interaction with the one seeking healing will be a little more involved. This is where the Five-Step Model below is so powerful, as it breaks down the different stages that tend to occur, helping ensure that you pray in the most beneficial ways possible.

In addition, as you seek to mature in effectiveness in healing (because, like any other spiritual gift, you can become more effective at healing the sick), you can use these five steps to review and assess where you need to grow further.

Finally, this is a framework to train those you lead and influence, so that you become a community of high faith, compassionate care, and best practice in healing the sick.

A COUPLE OF QUICK PRACTICAL TIPS

Before we dive into the Five-Step Model, a few very practical tips:

- Other than when just praying a quick prayer, we find it best to minister in pairs. When one of you is praying, the other can be listening to the Lord, and when they pray it gives you a chance to observe what the Spirit is doing.
- When in pairs, try to ensure that at least one of the people praying is the same sex as the person being prayed for, as that creates a safer space.
- As the one praying, you should always keep your eyes open, so you can observe what is going on in that person who is receiving prayer. You will learn to identify when the Spirit is moving upon a person - it can be helpful to gently speak that observation out loud, as it hugely builds faith. Occasionally the presence of God will fall so strongly on a person that they begin to wobble around or even look like falling over, and by having your eyes open you can make sure that they stay safe and at ease.
- Another tip is to have a box of Kleenex to hand. Deep emotions can be tied to illness, which often result in tears being shed. Additionally, tears can also be one of the signs that the Holy Spirit is at work.

FIVE-STEP MODEL

History

The Five-Step Model that you will learn here has its origins in the ministry of John Wimber, who led the Vineyard denomination through massive growth. Back in the 1980s and 1990s, Wimber traveled extensively, modeling how to operate in the power of the Holy Spirit. In particular, he was very effective at training people in how to heal the sick. The conferences he led combined intimate worship, practical biblical teaching, and highly accessible ministry times. Wimber wrote several excellent books, details of which are in *Next Steps 6 - Further Reading*.

Having played with this process for several decades, we have developed some of the language and added in our own insights and experiences. We would encourage you to engage with each of the steps in your thinking and practice not as an unbreakable formula (because it's not about a system), but rather as a proven and useful framework designed to help you better join in with what Jesus is doing in a specific individual.

The five steps are:

1. **LISTEN** — *Identify the Real Issue*

2. **DISCERN** — *"How should I pray?"*

3. **PRAY** — *Share the Father's Heart*

4. **REVIEW** — *"What's going on?"*

5. **CLARIFY** — *How to Move Forward*

———

1. Listen — Identity the Real Issue

On many occasions, before Jesus actually performed a healing, He asked the person "What do you want me to do for you?"

When someone is in need of healing, we can do the same! Simply start with a smile and gently ask, "What would you like prayer for?"

Do not assume that you know! We have had people hobble up on crutches, and we're all ready to pray for a leg healing, when they're seeking prayer for something entirely different.

As the person seeking healing shares, your task is to listen simultaneously in two parallel realms: one is the natural, the other is the supernatural. You must pay close attention to what the person is saying, and even closer attention to what God is saying. As Francis MacNutt puts it,

"Give one ear to the person and your best ear to God."

You might need to ask a few follow up questions, either if things aren't clear, or if you need more detail ("Where exactly does it hurt?"). The Spirit might lead you to ask questions around the depth and length of the issue, and maybe what else was going on in their life at the time this issue arose.

EMOTIONAL INTELLIGENCE

The *Listen* stage also sets the tone for the experience that the person who is seeking prayer will have.

Be welcoming and reassuring, since they might feel anxious about either their need (especially if it is private or embarrassing) or what will happen in the prayer time. **You need to model a kind and honoring posture towards them as an individual, so that a safe and sacred space is created for**

those few minutes. Don't forget that the number one goal is for that person to experience the Father's loving heart for them, regardless of whether or not they receive healing in that moment.

When someone has a major illness, you might find that they ask for some specific interim things - such as for the pain to diminish, the ability to hold their child again, or a good night's sleep. These are excellent things to pray for, since they can be measured quickly and immediately, and make a meaningful difference in their life. You can always ask if it's okay to pray for the big miracle (complete healing) as well if that's not mentioned. But these smaller steps along the way can build faith when someone has become worn down and discouraged.

Be aware that some people love to launch into a full medical history, with a blow-by-blow account of every procedure, appointment, and set-back. That is not necessary, and in our experience is usually unhelpful (and dull!). It rarely builds faith, and instead can feed into a tendency that some people have to gather as much sympathy as possible, rather than look to the Lord.

If you find yourself with someone like that, you might need to cut across them by saying something like, "Let me just interrupt you, because I want to honor your time, and also because I'm aware others are waiting for prayer [if that's true]. What we've found in praying for healing is that we need to lift our focus off the actual problem, and place our gaze onto Jesus, because He is far greater than any disease, situation, or human doctor. He is the One who can make us whole and healthy! So I think we should spend our time in prayer, but before we move into that, could you just tell me where exactly the pain is on your body..." [or insert another very specific question, if you genuinely need to know that information - otherwise, move into prayer!]

Build Faith

Make sure that all your words build faith in the person who has sought prayer. If they seem worn down, disappointed, or low in faith, you might want to share a 30 second teaching that sparks the gift of faith, preferably anchored in Scripture. And if you have seen (or heard about) someone with a similar condition being healed, definitely share that.

An example might be, "As we move to pray, I want to remind us that in the Gospels we see that Jesus loves to heal those who are battling an illness. So I have great confidence lifting you before Him today, because Jesus' heart towards you is so good and full of love. He hates this sickness more than you or I do, and He is far greater than this illness that the enemy is throwing at you. The Bible says that it's by Jesus' wounds we are healed, and so we are going to ask God to enforce that victory over sickness in your life today in a fresh way. Does that sound good?"

2. Discern — "How Should I Pray?"

This second step isn't one that the person being prayed for will be particularly aware of, since this has already begun in you during the *Listen* stage and will continue on into the actual praying.

For you, though, it is vital. In essence you are asking the Lord, "What sort of situation is this? And what sort of prayer or response is needed?" Put another way, you are discerning both what to pray for, as well as how to pray for it. This is because how you pray for a simple head cold will be different to healing an emotional wound, which in turn is distinct from a situation where demonic activity needs to be addressed. Alternatively, someone might ask for prayer for a stomach ulcer, when the underlying issue is actually stress from a difficult work environment.

The way into this discerning is to ask the Holy Spirit to show you what is going on and why. Gifts of the Spirit at work might include words of knowledge, words of wisdom, prophetic insight, discerning of spirits, tongues, or intercession. We write about each of these in other books in the Naturally Supernatural Series, but for now, don't forget that these are gifts that the Spirit can release to any believer at any time. They are gifts, not rewards or prizes, so don't be intimidated - you can ask for what you need at any time!

The bottom line is that throughout this stage, you are trying to hear Jesus on what is the ultimate cause of the problem that this person is facing. Why is it happening to the one asking for prayer? Bear in mind that sickness can originate from a mixture of natural and supernatural means.

Sometimes you have no idea what is going on! It's okay to be honest about that. If helpful, admit you're not sure ("I don't know why you have this disease"), but proclaim that the Lord does, so you're going to bring that person to Him and see what He wants to do. Don't forget to include your prayer partner in this stage - you should be asking them what they're sensing from God. That is one of the benefits of praying in teams of two or more people.

Be wary of falling into the trap of making assumptions about what is going on. This occurs when we forget that the same condition can be caused by different things, and thus will require different responses. For instance, you might have successfully prayed two weeks ago for a lady with a stiff shoulder, but don't assume that the stiff shoulder in front of you today is identical in its origins.

Our core dependency is upon Jesus. We must not simply repeat what worked last time, as that is placing our faith in previous experiences, rather than in the Spirit of God at work right now. We are to operate out of His current presence, not out of our past memory.

Different Types of Prayer

In *The Essential Guide to Healing*, Randy Clark summarizes what he's found to be the most common causes of illness, which in turn will impact the type of prayer that needs to be offered. Our experience would support that these are helpful categories through which different situations can be processed:

- **Natural Causes** — Ranging from accidental injuries, to a cold caught from a colleague, and even extending to a major illness directly caused by known environmental factors. Here the prayer is likely a straight forward pursuit of healing.

- **Psychosomatic Issues** (meaning an illness caused at least in part by mental or emotional factors) — Often you will find that the presenting illness will only leave once the person has dealt with the underlying issue. Examples of how to do this include repentance (if personal sin is in some way involved), forgiveness (whether of someone else or of themselves - this is a very common issue), submission of their past experiences to Christ, or perhaps a deliberate breaking of whatever has caused bondage in their mind or emotions.

- **Lifestyle Issues** — When someone ignores sound wisdom on how to treat their body (think of things like diet, rest, exercise, smoking, use of drugs, etc.), sickness can follow. If that is the case, then there will probably need to be confession and repentance, and perhaps a renunciation of where that particular thing has become an idol in their life.

- **Demonic Affliction** — As the Gospels make clear, there can be a link between illness and enemy oppression. This will require the afflicting spirit(s) to be commanded to leave. When dealing with an issue, it is often helpful to go back to what comic books call 'the origin story' - namely the narratives and experiences that opened the way for the enemy to bring oppression. It is in those root places that we ask Jesus to first bring cleansing and healing. Thereafter, the physical healing will often come far more easily. We will address this in greater detail in our future book in the Naturally Supernatural series on Deliverance.

- **Generational Curses** — The Scriptures have much to say about how blessings and curses can be passed down through the generations (the classic text being Deuteronomy 28). If the person seeking prayer talks about repeated patterns of sickness in a family, or of curses being spoken over them, then that needs to be renounced and its hold over the family broken.

It's Not a Magic Formula

Having given you some categories and suggested some different ways to pray, we want to urge you not to become hung up on the micro details of what to do and how to pray! The more you focus on obeying a formula, the less space you will have to spot what the Father is doing in the here and now.

Jesus said, *"Very truly I tell you, the Son can do nothing by himself; he can do only what he sees his Father doing, because what-*

ever the Father does the Son also does." (John 5:19) This is a vital principle when obeying His command to heal the sick.

We need to pay close attention to what God is up to in a specific person or situation, and then join in accordingly. As you move forward, you will find that while principles and patterns emerge, every person's need is unique!

3. Pray — Share the Father's Heart

When we pray with someone, our top aim is for that person to leave having experienced more of God's Father heart for them. That means they feel valued, encouraged, and loved, even if there was great challenge in the moment and spiritual battle took place.

Often those in long-term or major illness feel left out, over-looked, confused, or simply deeply discouraged. In such situations, the tone and texture of what and how we pray is often equally as important as our actual words. It might be that the Father's top agenda for them that day is that they know afresh how loved they are by Him, regardless of whether or not a physical healing takes place.

Of course, having said all that, we still want their healing, and so most of what we pray is focused on that taking place!

HELP THE RECIPIENT TO RECEIVE PRAYER

When we have gone through the (usually brief) *Listen* phase with the person, and as we move into *Pray*, you might want to prepare the person to 'receive' prayer. Some things you might do include:

- Encourage them to take a posture of receiving from God. We might say, "Some people find it helpful to close their eyes so they don't get distracted, and sometimes folks will hold out their hands as a sign of openness to Jesus. But just do what feels comfortable and helpful for you, there are no rules about any of this!" And of course these steps are not necessary or required - we just try to be attentive to the prompts of the Holy Spirit in these things.

- Request permission to place a hand on them (see note below).
- Ask them to let us know during the process if they sense anything from God, and say that we might stop and ask them what they are perceiving as we go along. This both prepares them for what's coming, and also helps them to pay better attention to those little nudges from the Spirit.
- Let them know at points that there might be a minute or two when we're silent and listening to God for them, so they should keep engaged with Jesus in the quiet. At a very practical level, we find that leaving our hand on the person while we are doing this acts as a non-verbal indicator that we're not done yet!
- Invite them to relax and simply enjoy God's loving presence.

LAYING ON OF HANDS

Importantly, we'll always ask if it's okay to gently place a hand on their shoulder, or on a specific place on their body that relates to their healing (e.g. their wrist for carpal tunnel syndrome, or the part of the back where they indicate there's pain). We might say, "Would you be comfortable with me gently placing a hand on your shoulder [or where the pain is in your back, etc.] as we pray?"

However, as with any physical touch, you must always be wise and sensitive. If the person is of the opposite sex, be very cautious. For instance, we make clear to eager young men that, even though they of course have the purest of motives, just because a sweet young woman wants to experience more of Jesus in her heart, they still should NEVER put their hand there on her body!!

As an aside, we strongly encourage you not to pray solo with a member of the opposite sex, as much as anything for the sake of appearance. At a practical level, if one of us is by ourself (say during a response time after a sermon) and sees someone of the opposite sex still waiting for prayer, we'll grab another team member of that sex to join us in praying. We've even said to the person waiting, "I'm just going to find another member of the team, as we like to pray in pairs," and they are always totally fine with that.

It's also important to remember that where people have experienced abuse or inappropriate touch in their past, they might find the laying on of hands to be a distraction or unhelpful, which is one reason we ask permission first. You must always honor their request, and never indicate disapproval or doubt at their choice.

Nevertheless, the laying on of hands is both biblical and powerful, and is a helpful thing to do in most cases.

BUILD FAITH

We mentioned this at the end of the section on *Listen*, but often we might begin our praying by verbalizing some faith-building declarations. A few examples might be, "Thank You Jesus that You love to heal the sick!" "We thank You for Your presence here." "Thank You that you love Taylor so much, and that your heart towards them is so wonderful and kind".

TONGUES

If you have the gift of Tongues, then this can be a good time to quietly exercise it (almost under your breath), as you seek to attune yourself to what the Father is doing and saying. We find that praying in the Spirit in this way helps us to discern

God better and is a helpful tool if there is any spiritual warfare going on. It is like our spiritual antennae go higher, we re-tune into Him, and then oftentimes breakthrough comes. But if you don't yet have this gift, don't worry, praying in English or another human language also works really well as you tune into the Spirit.

Pray Short but Sweet Prayers!

As a general rule, prayer for healing does not need to be an epic speech that does two laps of the planet and covers every possible nuance or potential outcome. When we pray, we are not legally obliged to list all the terms and conditions around what is taking place!

Other times you might need to spend longer - perhaps 5 or 10 minutes for something more weighty - as you discern the heart of the Father for this person and situation, and then pray accordingly. We find the need for additional time to be especially true for issues that are tied to the demonic or emotional healing.

However long you spend, seek to command sickness to leave.

Exercise a humble but godly authority as a child of God.

Speak in a manner that leaves the person feeling loved and built up in faith.

We might pray something like, "Father, please come on this person and release Your healing power." "Holy Spirit, come!" "Lord Jesus, please come and touch his body and bring complete healing." "Arm, be healed in Jesus' name!" "Headache, be gone!" "Cancer, we curse you in Jesus' mighty name!"

Then we wait and watch.

By the way, when you pray, speak in your normal voice! Don't switch into some strange religious tone. And you don't need to shout. Use your voice like you would in any regular conversation.

PHYSICAL MANIFESTATIONS

As we pause, we pay attention to what the Spirit is doing. As part of this waiting, sometimes there might be physical manifestations of the Spirit's presence, especially if you are spending a little longer in prayer. However, please note that this is NOT a universal rule (as a rough guide, we find this takes place in some observable way in perhaps up to half of all people).

If it does occur, there might be fluttering of their eyelids, tears (gentle or violent), smiling or even laughter, heat in the part of the body that needs healing, change in skin color or tone, physical trembling or even rocking, physical relaxation of the body as the peace of God falls, or simply a more peaceful look on their face. Sometimes you might notice a troubled look as you say something, which is something you need to note.

Having listed those examples, we should warn you that some of those will be from the Lord, some from the flesh, and occasionally some will be from the enemy. The flesh aspect isn't necessarily sinful - usually it is born out of a desperation and desire for more of God, or for His healing. Very occasionally you might discern that a person is seeking attention by acting strangely, but that doesn't automatically mean you need to call that out.

We should not be surprised by a demonic manifestation, since when the light comes, the darkness has to flee! But those signs act as indicators for us about how we are to pray. Things you might see could include grimacing or snarling as Jesus is mentioned or His presence invoked, or the sense that prayer for healing is being rejected by the person or something in the person. Again, the spiritual gift of discerning of spirits is very useful here, so quietly ask Jesus for His insight if something seems off to you ("Lord Jesus, would you reveal now the spiritual source of what is going on here").

If there is something demonic that seems to manifest, or you sense its presence, or if the person says they wonder if that's the case, don't freak out! The simple way to deal with it is to command it to be gone in Jesus' name. You are simply exercising the same authority that you do when you minister healing. If it makes a noise, just tell it to be silent. If it still hangs around, you can bind it in Jesus' name and then calmly seek help from someone more experienced in these things.

The other comment about physical manifestations is that we are not to chase them. Never evaluate the success or otherwise of a time of prayer ministry by them. And absolutely never suggest or hint to a person that they need to act in a certain way in order to demonstrate that they are encountering God.

Prophetic Sensitivity

When we are healing the sick, it's common for the prophetic gifts to flow as well. **Even if you don't think of yourself as particularly gifted in this way, the Spirit might come upon you with specific guidance for that moment.** Often this might feel like it appeared out of the blue, and you'll kind of know that it didn't originate in you.

For instance, the Holy Spirit might release specific instruction about how to pray, or insight into what the root cause is that lies behind the illness. There could be a word of knowledge (information that you otherwise couldn't possibly know in the natural), or a word of wisdom (perhaps direction on how to pray to unlock a situation). You might sense that the person needs to repent of something, or offer forgiveness, or renounce something where the enemy has a foothold. Other times you might be stirred to pray in a certain way, even with a specific action accompanying it (such as symbolically cutting off ties to an unhealthy part of their family line).

Your task is to go with what Jesus is saying and asking you to do. We'll be honest here - this can often be (very) nerve wracking! However, we've also found that if we simply ignore that prompting, then that anointing may well lift and we'll have missed the moment to partner with God.

As always with prophetic gifts, share any words with sensitivity and love, recognizing that you might be wrong. Never speak them out as definitive - always invite the person to weigh it for themselves.

We might say something like, "Max, as we're praying, I am sensing that the Lord might be saying this to you…." Or, "I'm wondering if some of what is going on here in this illness is coming from the enemy - does that resonate in any way with you?" Or, "You mentioned earlier that others in your family suffer from similar conditions, and I wonder how you'd feel about renouncing and breaking off any curse that in the past might have been placed upon your family."

As you read these examples, bear in mind that the problem with writing these things down is that they can sound kind of strange out of context! The key is to keep humbly attentive to what you sense is the Spirit's leading, and to then follow Him in obedience and faith.

Don't Counsel

Finally, resist the temptation to turn prayer ministry into a counseling or advice session, and certainly don't preach! Yes, those are indeed legitimate gifts that should be appropriately exercised within the church, but this is not that time. This can be hard to control when you can see how they ought to be living, but please exercise the fruit of the Spirit that is self-control.

Be especially careful not to assign to God what are actually just your thoughts, sanctified or not. Bear in mind that it might be appropriate/ possible to share one or two of your thoughts in Step 5, *Clarify*, and we'll explain how to do that when we get there.

Generally...

For as long as God is touching the person, continue working with Him. Give thanks whenever you see any progress. **Sometimes the Lord will give you plenty to pray out loud, other times it is more a ministry of presence that you are offering**, and on other occasions it will be a brief prayer and then you're done. Again, this is something that you can only discern in the situation (yet another reason to pray in pairs!).

When do you stop praying?

- When the person being prayed for lets you know they've finished.
- The Spirit tells you to stop. Usually this means the person is also finished, but sometimes they might still be in prayer and you leave them in the care of the Spirit to enjoy God's presence.
- You run out of things to say and pray! That happens, and it's okay.

4. Review — "What's Going On?"

After praying for healing with a person, we will pause to ask them what's going on. The reason we ask is because, even though we have the gifts of the Spirit, we're not mind-readers! Also, it builds a sense of shared responsibility for the process.

The aim of *Review* is to find out more about what the Father is doing in this situation. Often the person will feel something before you see it, which is why it's important for them to tell you.

We find it works best to pivot into this dialogue in a gentle way, so that it almost flows naturally from the praying and waiting on God. By way of a very granular practical tip, we don't do a big 'Amen' ending to the prayer before this, so that the person is more likely to stay in a place of engagement with God.

Often Christians don't like to do this stage! After all, we're actually asking for proof of what we're praying, which feels hugely risky. In that moment, we think to ourselves, "What if nothing has gone on? Does that make me a failure? How do I defend God's reputation?" And so the easy option is simply to say 'Amen' and move on from the prayer, without stopping to check in and review with the person who is receiving prayer.

JESUS DID THIS!

And what a great encouragement that is! In Mark 8:22-26, Jesus is called upon to heal a blind man who lived in Bethesda. After initially praying, we read: *"Jesus asked, 'Do you see anything?' [The man] looked up and said, 'I see people; they look like trees walking around.' Once more Jesus put his hands on the man's eyes. Then his eyes were opened, his sight was restored, and he saw everything clearly."*

Following Jesus' modeling, if what we are praying for is a physical condition that can be tested, then we suggest that the person does so there and then. Usually that person will pause and look at us like we are slightly crazy! So we then reassure them that we are praying to see results, and encourage them to test the condition.

It's so fun to watch someone test out their previously frozen shoulder/ unbendable back/ broken bone etc, and to see their eyes come out on stalks as they realize that they can now do things that a minute previously they couldn't even attempt. God is so good!

HEALING CAN COME IN STAGES

Sometimes we find that the person is partially healed at this testing stage. In that case, we will encourage them that they have experienced some healing and suggest that we pray some more for complete healing. People often won't have a grid for healing that takes place in stages, so we find it helpful to reassure them that this is perfectly normal and often occurs.

We then go back to stage three, *Pray*, and see what happens.

Alternatively, with conditions that can't be tested there and then, the person might have sensed something from God that really helps in lasering in our prayers. This might be insight into the root cause of the condition, or even greater clarity about the exact diagnosis, or perhaps a picture about what Jesus is doing in them. As they share this, it can feel completely appropriate to pray again for their healing, as you respond to this revelation. By way of example, we have had people tell us that while praying they have been reminded of a situation from their past, and wonder if it might be connected to their presenting issue, which of course is a huge indicator for how you are to pray next.

This pausing to review will come more quickly when doing a quick prayer, or when offering healing to the lost outside of a church service. For instance, if we pray with a lost neighbor for healing, the actual praying time might only be 15-30 seconds, so the *Review* stage is quickly reached. But we always try to ask them what is going on, and if they can feel any difference.

5. Clarify — How To Move Forward

Once you have finished praying and reviewing, a good way to wrap things up is to help that person clarify their next steps. The depth of this will probably depend upon the length and weightiness of what you have been praying into.

As with the other stages of the Five-Step Model, it is important to be attentive to the Spirit, and not simply trot out the same thing to each person. However, in the absence of particular direction from God, you'll find yourself returning to certain pieces of wisdom.

- **If they've been healed:** Encourage them to share the testimony THAT DAY. We might ask, "Who do you know who will be pleased to hear this, and will be excited about what Jesus is doing in your life?"

- **If they've been partially healed:** Invite them to keep on thanking God and asking Him to complete the good work that He has begun in them. We have seen many instances where healing is a process that takes place over the hours that follow (and often overnight). "As you go through the rest of your day, keep on thanking Jesus for what He's done, and ask Him to give you complete healing. This is a good thing to pray as you lay in bed tonight!"

- **If there is a lifestyle issue:** Ask that person what their specific next step is in making changes, and who will hold them accountable for that. (By the way, it is usually better for someone else to help them in this way, unless you are already especially close to them.)

- **If there was an afflicting spirit:** Make sure that person is aware it might try to return, and equip them in how to rebuke it in Jesus' name. "Sometimes the enemy will try to return. If you sense something dark later on, simply say, 'Be gone in Jesus' name!', and begin to praise God for who He is and thank Him for what He's doing in you."

- **If they were not healed:** Warmly invite them to come back for prayer as many times as is necessary, and to choose the path of persistent faith. Make sure they do not listen to the enemy's lies (e.g. they are not worthy, it's their fault, God doesn't care, they are a bother to others, etc). Never blame them for a lack of faith, but do be honest about the situation ("Healing is a mysterious thing, and sometimes we simply do not know why things don't happen as we would wish"). To help them keep believing and experiencing that Jesus is at work, encourage them to be active with a church group where they are known and loved.

A few other pointers:

- If the person believes that they've been healed and has been on medication for that condition, NEVER tell them to stop taking it. Instead, they should go and seek advice from their medical professionals. Encourage them that this is also an opportunity to bear witness to what Jesus is doing!

- Sometimes the Holy Spirit will give you a specific prophetic direction, for instance asking the person to go and make peace with someone. Obviously share this with grace and humility (you might be wrong), and see how they respond. Likewise, they might hear some specific instruction from the Spirit as to next steps and want to process that with you. And it will amaze you how often the Spirit will speak separately to you and to them about the same thing!

- You might have some general counsel that is from you, rather than the Spirit. It's fine to share this, so long as you are not lecturing, being a bossy-boots, riding a hobby horse, or undoing all the good work of God by being judgy or condemning. Rather than download all of your extraordinary wisdom in one go, perhaps focus instead on one nugget to offer - "Have you considered chatting with Sandy about this? I know they went through something similar", or "We have a support group at church that might be perfect for you as you take your next steps", or "Who would be a good choice to help you walk through the next stages of this wonderful work that God is doing in you". Resist the temptation to reorganize their life for them.

- It might be helpful to encourage the person to go and do some specific Bible study on a topic related to their situation. You might ask the Lord for one or two passages to suggest as a starting point.

- Finally, you can't over-value testimony! We as believers need to make it a way of life. It raises the level of faith in the wider church and community when the miraculous stories of God are recorded and recounted. In the Old Testament, whenever Israel forgot what God had done they fell into decline (see Psalm 78:40-43). In Revelation 12:11 we read that Christians will triumph over satan by *"by the blood of the Lamb and by the word of their testimony"*. So please encourage anyone who is healed to tell others, and to be willing to write down their story / record a video on their phone / generally spread it to others. It really does make a difference!

Post-Praying

After you have finished praying for the sick, and you are no longer with the ones you prayed for, it's wise to briefly pray together as team members (if in the context of a church service), or with another believer (if this happened in the normal run of life).

- Start by thanking God for what He has done, and praising Him for His goodness, and for the honor of being used by Him to extend His Kingdom.
- Ask Jesus to cleanse you from any residue of the enemy. This is especially important if any sort of deliverance took place, however low-key.
- Ask for protection upon yourselves, your families, and your households, with no room for counter-attack from the enemy.
- Bless those you have prayed with, that they may walk closer with God as a result of the time of prayer, that any healings will last, and that those who weren't healed will feel deeply loved and ready for more prayer in the future.
- End by asking God to seal the good work that He has done, and to build on it going forward.

EXERCISE

This Five-Step Model may sound a little complicated and only for those who are mature in their faith - but that's not the case!

We have gone into great detail in all of the Steps to try to equip you for a variety of different scenarios. However, don't worry if you can't remember it all, or think you can't do it.

Remember: Jesus tells all of his followers to heal the sick. If we step out in obedience and remember that He who calls us is faithful, and that He is the One who will work through us, you'll be amazed at what Jesus will accomplish through you!

Now Do This...

Review the Five-Step Model. Here are the headings again to help you:

1. **LISTEN** — *Identify the Real Issue*

2. **DISCERN** — *"How should I pray?"*

3. **PRAY** — *Share the Father's Heart*

4. **REVIEW** — *"What's going on?"*

5. **CLARIFY** — *How to Move Forward*

———

Consider:

- Which step comes to you most instinctively? Which one is the least natural for you?
- In your own words, write down the essence of each step. What is the key thing you need to remember?
- How can you become more proficient in each step? Consider a specific action that will help you embrace

the strength of that stage in the process, especially if that is not something you would normally do.

- Write down the Five-Step Model in a place where it will be handy (we'd suggest as a note on your phone), and bring it out when you have the opportunity to pray for healing in a church group or service. Ideally you should memorize this as a template from which you can work.

The great test
of obedience
comes in the
darkness,
rather than
the light.

WHEN THE HEALING DOESN'T COME

B ack when we were dating and then first married, we became good friends with the couple who had originally planted Hannah's home church. Very sadly, about 10 years previously, Dave had been diagnosed with Parkinson's disease, and could no longer work. Yet he and his wife Sue had beautiful spirits and were deeply in love with Jesus.

Four years into our married life, we ended up moving to that church, and shortly thereafter leading it. Throughout our time there, Dave and Sue were such encouragers and supporters, and it was always a real joy to go and spend time with them.

Many church members had prayed and fasted for Dave over the years, both privately and as part of church-wide days of prayer. It was quite a well-known church, and a lot of visiting preachers and leaders would pass through and pray for him, but tragically he was never healed of the disease.

At one point in our time there, Alex felt led to do a sermon series on the gifts of the Spirit, including healing. We had some anxiety about this plan; obviously, Dave's lack of healing would be the elephant in the room, so we went to see him in advance. The amazing thing was how adamant he was that there should be bold biblical teaching on healing, that the church should be encouraged to step further into this gift, and people should be equipped to pray for the sick. Here was their founding pastor, 15+ years into this horrible disease, urging that we should in no way soft-pedal on his behalf around the issue of healing.

He was clear that even though he hadn't received healing, this was still the right thing to teach on, because the Bible commands us to heal the sick.

Dave and Sue had every excuse in the world to become bitter, or say that this was dodgy theology due to his experience, but they refused to go there. Instead, they rooted themselves on the clear teaching of Scripture and their commitment to Jesus.

Very sadly Dave died of complications from the disease about six years later, and did so as an incredible hero of the Kingdom.

Nonetheless, this raises many difficult questions. How do we process such tragic losses? Did Dave not deserve the healing enough? Was his approach naive? How do we handle such situations both pastorally as well as theologically? Can we find a place where those who aren't healed still feel fully loved and valued, while we simultaneously operate as a people of faith who expect to see healings take place around us?

THE MYSTERY OF HEALING

There is an aspect of healing that remains a deep mystery to us, even after decades of seeing the sick healed in response to prayer and faith in Jesus. Just why is it that some are healed, while others are not?

We have been in rooms where a number of people are suffering from the same ailment, yet with the same people doing the praying, only some are healed. We've learned that we can't pin the missing element upon the one who is sick, since we have watched godly, humble believers with what seems the perfect attitude not be healed, while a hugely immature person with flaky beliefs walks away whole. There have been times when we've prayed once for a stranger with terminal cancer and they've been instantly healed, while others we love dearly, for whom many have fasted and sought the Lord, have died.

There is something simultaneously wondrous and yet deeply frustrating about the whole enterprise of healing the sick.

But because the Lord Jesus calls every disciple of His to this task, the one thing that we can't do is to give up trying. One of the reasons that so much of the Western church has neglected praying for healing is due to the pain of when it doesn't happen. Yet part of following Jesus is that we commit to always obeying His commands, even when they don't make sense to us or we don't see the results we desire.

The great test of obedience comes in the darkness, rather than the light.

This chapter, therefore, is an attempt to face up to the theological and pastoral questions that arise when healing doesn't come as we would wish.

In seeking to give you some pathways through, please would you be patient with us? We aren't able to sit across from you and hear your stories, hence we're unable to temper our responses in ways that are sensitive to your own personal disappointments. So if at any point what we write comes across as insensitive, please forgive us - that is the very opposite of our heart.

At the same time, though, we do not want to have to qualify every statement (which would quickly make for an extremely long and dull read!). This means that at points we will cut to the chase without stopping to write out every possible nuance. Again, that isn't because we don't care, but merely for the sake of brevity.

Finally, we also want to make clear that we don't think we've discovered some ultimate 'system' to explain away all the objections and difficulties. Earlier in this book we spent time on biblical principles, and in a 'perfect' theology it is true that God always heals (in other words, that is our heavenly hope). But today we live in a fallen world, where Jesus might have all authority, but He's delegated that to us. We operate with a 'tarnished' theology, where we simply can't fully explain why things don't quite happen as we believe they should.

A colleague of ours was once preaching on why bad things happen to good people, and he built a huge jigsaw to illustrate his talk. As he introduced a new main point of explanation, he inserted an additional piece into the puzzle, each time bringing it closer to completion. However, as his talk neared its end, there remained a large piece missing in the middle of the puzzle. He explained that this was deliberate, since this side of heaven none of us can completely explain why life at times seems to be so unfair.

In the same spirit, we are going to offer you some direction and insights that we have gathered over the years, and hopefully you will find most of them to be thought-provoking and helpful.

We are also deeply aware that some people — maybe you or a close friend — are holding in tension a deep trust in the goodness and healing mandate of God, even while they or loved ones fight long-term or major illness. We honor your faithfulness, faith, and love, and pray that these words might be some small part of the Lord's healing balm for your heart.

NEW TESTAMENT EXAMPLES OF PEOPLE NOT BECOMING WELL

Given the remarkable healing ministry of Jesus, and the many healings performed by the early church, it might come as a surprise that the New Testament records several instances of people (seemingly) not being healed.

Jesus did heal everyone who came (or was brought) to him. There is no example of Jesus turning away anyone, refusing to heal, or failing to heal. It is fair to say that Jesus loved to heal the sick! **However, Jesus did not heal everyone who was sick in His time.**

The only occasion where we're told Jesus struggled with healing was in His hometown, and that was due to lack of faith (Mark 6:1-6). Did He try and fail, or did He just see there wasn't enough faith so didn't even try? What made the individual healings He could do successful?

At the pool of Bethesda in John 5, which in some ways was the cultural equivalent of a primitive hospital since the sick gathered there, the text implies that only one person was healed by Jesus on that occasion (although we can't be certain - see John 5:2-9). The key to understanding comes a few verses later, in John 5:19, where Jesus says of Himself, *"the Son can do nothing by himself; he can do only what he sees his Father doing, because whatever the Father does the Son also does."*

During His earthly ministry, Jesus, while remaining fully divine, performed supernatural works out of His humanity under the empowering of the Holy Spirit. He operated in such close partnership with the Father and the Spirit that He only did what He first saw Father doing. This creates the model for how each of us can approach healing the sick.

The other story about Jesus seeming to limit healing comes when He is approached by the woman born in Syrian

Phoenicia (Mark 7:24-30). This place of origin is lost to us as readers of the Bible today, but to the original hearers they know it meant the woman was not only not a Jew, but also she was from a nation openly hostile to God and to His people.

The dialogue between her and Jesus is remarkable culturally, since both of them break the strict social rules around interactions between men and women, and Jews and Gentiles. However, it is also quite playful — imagine Jesus with a twinkle in His eye, as He calls forth the woman's faith. She responds in a similarly hyperbolic manner, and we quickly see that this is not a negative view on healing that is being presented. Possibly the issue in this encounter was one of timing, as this took place prior to the commissioning of the church to go into all the world - yet even then Jesus operated with joyful grace and flexibility.

Moving into the writings of Paul, four instances are interesting to consider.

- In Philippians 2:25-27 we're told that Epaphroditus *"was ill, and almost died"*. It seems reasonable to assume that Paul and other disciples prayed for healing, but clearly that didn't occur promptly and thus he came very close to death. However, he didn't die from that condition, so either the disease ran its course or he was healed later on in the process.

- Paul has a deep fatherly concern for Timothy, and in 1 Timothy 5:23 he instructs him, *"Stop drinking only water, and use a little wine because of your stomach and your frequent illnesses."* Presumably Paul and others had prayed for Timothy's delicate digestive system, but he remained a young man susceptible to illness.

- In 2 Timothy 4:20, an obviously lonely Paul writes that *"I left Trophimus sick in Miletus"*. This wasn't something he wanted, so it seems fair to read into this text that prayer for healing hadn't worked. There is some discussion around the specific Greek phrase that Paul uses about Trophimus, which implies overwork as the root cause — perhaps this man had overworked and was not taking proper care of himself. Whatever the exact meaning, at the time of writing Trophimus had not received healing / restoration in a supernatural way.

- Paul himself wrote about his own experience of sickness. In Galatians 4:13-14 he recalls, *"As you know, it was because of an illness that I first preached the gospel to you, and even though my illness was a trial to you, you did not treat me with contempt or scorn."* Scholars tell us that Galatians was one of Paul's earlier letters, and this illness is not mentioned in any other correspondence that we have from Paul, so he either recovered or was healed. While Paul is careful not to assign the illness to God's hand, he does recognize that the Lord can use sickness for His purposes (see Romans 8:28).

"BUT CAN'T GOD USE SICKNESS TO DISCIPLE US?"

Sometimes we'll be asked a question that runs along these lines: "There are times when God chooses, in His sovereignty, not to heal some who are sick. Obviously, in those cases, it was not His will to heal. I've seen God use sickness in peoples' lives to bring them to the place of repentance and faith and turn to God for their salvation. God can and does sometimes use sickness for the ultimate, albeit not temporal, good."

Clearly God can use sickness to work in someone's life (for instance, to bring them to repentance). Likewise, God can use a person's sin to shape them, even though the sin is neither from Him nor is it His desire for that person.

Simply looking at the life and teaching of Jesus, and how He responded to every sick person He met, we conclude that God does not desire for us to be in either sin or sickness.

Therefore, as a general principle, we should always battle both, even if we don't immediately receive the outcome we desire for our family and friends, because Jesus wants neither to be our ongoing experience.

If you baulk at this sentiment, it could be that you are attaching a false value to suffering. While occasionally suffering might have some redemptive aspect to it, the weight of Scripture and life experience show that sickness rarely has this positive impact. Almost always sickness is a negative thing (hence unhappiness and even depression are tied to ongoing ill health). While throughout history humans have valiantly and admirably sought to make the best out of a difficult time, it is an entirely different thing to say that this is God's Plan A for us.

WHAT ABOUT PAUL'S 'THORN IN THE FLESH'?

When we share our belief that God does not give sickness, even to improve our character, someone will respond, 'But what about Paul's thorn in the flesh? Wasn't that a sickness that God refused to remove, in order to deepen his character and trust in God?'

Paul wrote about what he called his *"thorn in the flesh"* in 2 Corinthians 12. He began the chapter by saying that he had been receiving *"surpassingly great revelations"* from God. These were downloads from God that were remarkable in quantity and quality - he was, for instance, permitted to see into the third heaven, where God is enthroned in majesty and glory.

One of the results of these revelations was that Paul came into a place of tremendous spiritual authority and power. This resulted in considerable attack and hostility from the enemy. From the previous chapter we read that this resistance came in a variety of forms.

> *"I have been constantly on the move. I have been in danger from rivers, in danger from bandits, in danger from my fellow Jews, in danger from Gentiles; in danger in the city, in danger in the country, in danger at sea; and in danger from false believers. I have labored and toiled and have often gone without sleep; I have known hunger and thirst and have often gone without food; I have been cold and naked."* (2 Corinthians 11:26-27).

It is important to note that the way Paul describes it, the trials he faced were predominantly relational in form, whether through direct opposition (violence, prison, etc.), or indirect hostility that led him to live in extremely challenging circumstances to avoid arrest or attack. It is in that light that we can best interpret the phrase in 2 Corinthians 12:7, where he writes that *"in order to keep me from becoming conceited, I was given a thorn in my flesh, a messenger of Satan, to torment me."*

Some say that he is talking about an illness, due to use of the word 'flesh'. This then leads to the view that God sometimes chooses not to heal the sick, due to the following two verses: *"Three times I pleaded with the Lord to take it away from me. But he said to me, 'My grace is sufficient for you, for my power is made perfect in weakness.' Therefore I will boast all the more gladly about my weaknesses, so that Christ's power may rest on me."* (v.8-9) It is suggested that the use of the word 'flesh' means that this was a physical ailment. However, this is not backed up by the context, where Paul focuses not on ill health, but rather demonically inspired opposition.

He explicitly states that this thorn was *"a messenger of Satan"* — in other words, it was demonic opposition made manifest through human behavior, that brought the enemy's message in direct resistance to the message of the Gospel. These opposing messengers were people, mostly outside but even some inside the church, who planned and worked to attack Paul and his ministry.

Paul goes on to share the words of grace that Jesus spoke to him, which include the concept of God's power being revealed through our "weaknesses" (another way of describing "thorns in the flesh"). In verse 10 he reflects on the nature of this weakness, writing, *"That is why, for Christ's sake, I delight in weaknesses, in insults, in hardships, in persecutions, in difficulties. For when I am weak, then I am strong."* Notice that Paul describes the weaknesses (thorns in the flesh) that God allows in purely relational terms. There is absolutely nothing here about sickness.

One of the principles for interpreting the Bible is to look for other occasions where that language is used. As a well-educated Jewish scholar, Paul would have been fully aware that *"thorn in the flesh"* (or side) appears in Joshua 23:13 and Ezekiel 28:24. On both occasions it is in reference to neighboring tribes who were opposed to the work of God, and who sought to pull God's people away from Him. In neither instance does the text even hint at health issues - instead, believers are warned away from relationships that will cause them to dishonor God.

So the *"thorn in the flesh"* text is fascinating, important, and meaningful, but is not anything to do with God withholding healing. Instead, it is a metaphor to describe hostile relationships, often with undercurrents of spiritual warfare, that seek to undermine our faith and witness. Thankfully God can also redeem these relationships, in order to better display His power.

TWO TYPES OF RESPONSE

The more you pray for the sick, the more you will see healings take place, which is an incredible joy and privilege! Yet, simultaneously, the more you pray for people, the more you will see some not being healed (at that time, at least). Tragically, there will be a few people that you pray for who end up dying of that illness or injury.

At some point, someone will ask you, "Why wasn't I (or my dad/ child/ friend) healed?" If you're anything like us, you will feel a sudden weight press on you as you recognize the seriousness of what they're saying, and the importance of what comes out of your mouth over the next minute.

In these moments, it is vital to identify what sort of question is being asked, in order to determine whether the response they're looking for is more pastoral or theological in nature.

The Pastoral Response

If the conversation comes in the midst of heartache, hospitals, and hopelessness, you must understand that what that person most needs to hear is the loving heart of a shepherd. They are facing up to loss, disappointment, frustration, and heartache. Even if they don't realize it, they're not actually asking you for a theological explanation. Instead, their greatest need is to experience the compassion, kindness, and love of God, ministered through you.

Sometimes this means looking past the actual words that are coming out of their mouth. "Why would God allow this?" is not always an invitation into a detailed debate about the nature of suffering in a fallen world. Instead, recognize that their question might instead be more of an expression of pain and a cry for help. Your response should be dictated by that deeper reality.

There are occasions when the most biblical thing to do is to wrap someone in a compassionate hug, and simply repeat, "I'm so very sorry."

At that point in time, the most important thing this person needs to know is that they're loved and not forgotten. Through your words and actions of kindness and care, you will also bring the love of Jesus into that situation, even if in that moment the person is furious with God.

If appropriate - and use great sensitivity here - it can be good to ask some open questions that might help the person identify where God is present with them. Examples might be, "Were you aware of God's presence? How did Jesus meet you in the pain? What did you sense in your spirit? Has your perspective changed on your situation?" Sometimes it's in the darkest moments that God seems extra present.

As an aside, where there is a lot of pain, someone might speak out some tough, even awful, things about God and how they feel about Him. If that's the case, don't feel the need to become Jesus' defense attorney - He's more than capable of fulfilling that role Himself! If you're offering a pastoral response, simply shepherd and love them well, and resist the urge to correct every piece of angry theology they vent. The wisest thing Job's friends did was to listen for a long time. It was only once they started talking that they messed up! Instead of debating, you can simply pray with and for them, and ask for God's loving care to flood around them. His character is still true and dependable, and we've seen Him answer many such prayers with His loving presence.

As time progresses, your presence and care in the crisis might open up the way for more substantive theological conversation, but never neglect or underestimate the importance of the pastoral response.

One way to spot the difference is by sensing whether this is wartime or peacetime for the person asking. In wartime, where the spiritual battle is fiercest and lives can literally be on the line, the pastoral response is what's needed. A pastoral response builds up the individual emotionally and spiritually, strengthening and encouraging them to keep in the fight. Peacetime, by contrast, is usually well away from the sick bed and emergency room. Instead, it's a time when a more reflective and nuanced conversation can be had and abstract principles processed without causing emotional distress.

The Theological Response

Separate from the intense crisis moment where the pastoral response is what's required, there will be other occasions when you'll rightly enter into a discussion about why sickness happens, especially in light of our claim that Jesus is a God of love.

Very rarely is this what is required at the hospital bedside or with a grieving relative. However, there will be plenty of other, more appropriate contexts for this conversation to take place.

We have already discussed the nature of healing earlier in this book, but by way of summary, we find our conversations tend to go back to a few main points:

- **God is good and full of love.** This means that He does not operate in a capricious, unkind, or cruel manner. Therefore we can confidently ask Him, our loving Father, for good gifts such as healing, because He loves to give us such things.

- **Sickness is not part of God's eternal heavenly Kingdom.** This is the same Kingdom that we're commanded to pray will come daily around us here on earth. While He can use anything for our good, this does not mean sickness comes from God. Therefore we boldly seek to attack sickness by ministering supernatural healing.

- **The Kingdom is both now and not-yet.** Jesus' ministry was built upon His bringing the loving kingly rule of God into people's lives. His death on the cross and resurrection sealed that great victory, and so God's dynamic activity on earth is now permanently present and visible. But we still have a wicked and cruel enemy, who still has some influence in this world until Jesus returns. Specifically, sometimes this means that we'll see amazing supernatural healing occur (the Kingdom is now), while other times we'll be left frustrated, even heart-broken, when there isn't healing (the Kingdom is not-yet). Candidly, we rarely know why the not-yet has come to the fore. However, recognizing that both realities (the now and the not-yet) can simultaneously be true provides a helpful framework for theological understanding. We can honor the greater truth that Jesus is at work, His Kingdom is advancing, and His followers are called to join Him in that endeavor, which includes operating in the gift of healing. At the same time, we have a way of understanding why our prayers aren't always answered and the enemy can seem to gain a victory, without turning God into some sort of moral monster who dwells high up on Mount Olympus.

- **We always have an eternal hope.** Even when the healing doesn't come, we maintain our longing for our reunion with God and our entering into the fullness of Jesus' Kingdom rule. We look forward to either our joining Him in heaven, or His return to earth and the making of all things new, including the final destruction of sickness, sin, and death. This doesn't mean we don't mourn those who die, or pretend there's no struggle with unanswered prayer, but those things are simultaneously paired with this greater hope.

REASONS WHY PEOPLE AREN'T HEALED

The Bible speaks to some of the reasons as to why healing doesn't always occur.

As one who ministers healing, this list can be a helpful tool for reflection, especially if there is a sickness where you're not seeing breakthrough. Prayerfully look through this list, and if the Spirit indicates something to you as being pertinent in a situation, consider your next steps. We have made a few suggestions each time under the 'Application' heading. Bear in mind that there could be a reason that is not listed here.

Obviously beware of making this a new legalism! Also be pastorally sensitive: even if your analysis is correct, someone in a heart-breaking situation might hear this list as a word of condemnation that induces guilt, which is a horrible thing for them to experience. So be wise - it might be that this is an internal process that you don't share out loud.

Nevertheless, reflecting on these hindrances can be a way to allow God to disciple you further in fruitfully exercising the gift of healing.

1. Lack of Faith

Jesus gave two main reasons why healing doesn't happen - one of which is lack of faith.

In Matthew 17:14-20, Jesus is asked to heal a boy with an evil spirit, after the disciples tried and failed. When the disciples ask why they couldn't heal, Jesus said, *"Because you have so little faith. Truly I tell you, if you have faith as small as a mustard seed, you can say to this mountain, 'Move from here to there,' and it will move. Nothing will be impossible for you."*

Jesus makes clear that He is exasperated by their lack of faith (v.17). He seems to suggest that the disciples should have had enough faith by this stage to do this act.

Of course, there are numerous situations where Jesus heals and deliberately draws attention to the faith of either the one who is sick, or their family member or friend. Examples include the woman who has had 12 years of bleeding (*"Daughter, your faith has healed you. Go in peace and be freed from your suffering."* Mark 5:34), the Canaanite woman from Syro-Phoenicia (*"'Woman, you have great faith! Your request is granted.' And her daughter was healed at that moment."* Matthew 15:28), and when Jairus is urged to have faith for the healing of his daughter (*"Jesus said to Jairus, 'Don't be afraid; just believe, and she will be healed'."* Luke 8:50)

We can therefore fairly conclude that lack of faith is a major contributor to lack of healing. Francis McNutt (whose book, simply called *Healing*, is excellent) thinks that this is the number one reason that healings don't always happen, and we are inclined to agree with him.

And for every single one of us, faith is an area where we can (need) to grow more, so that we can be used more by God.

APPLICATION: If you're not seeing healing, ask God if lack of faith is an issue.

If you sense that to be the case, repent, and ask Him for a fresh release of the gift of faith upon the one who is sick and those ministering. Reading out a few Scriptures about healing can be atmosphere changing, as can relevant testimonies of healing. You might also need to remove from the room those who are more content with dwelling in doubt rather than walking by faith.

2. Lack of Persistence

The other major reason that Jesus suggests for healings not occurring is our lack of persistence.

Some people assume that if God heals it will always be instantaneous, and so when that doesn't happen they quickly stop praying. While we see many immediate healings, there are other times when healing seems to be a process, requiring us to persevere and fight.

Jesus talked about this principle when urging us to be like the persistent widow (Luke 18:1-8) or the neighbor banging on the door at midnight to feed his unexpected guest (Luke 11:5-8). **We need to be the ones with *"shameless audacity"* (Luke 11:8), who persist and persevere in pursuing God for what we need, even when our situation seems dire.**

In Matthew 7:11 Jesus promises that, *"If you, then, though you are evil, know how to give good gifts to your children, how much more will your Father in heaven give good gifts to those who ask him!"* Healing is one of the good gifts of the Holy Spirit (see Luke 11:13) for which we are invited to persist in prayer. We pursue the now of the Kingdom to overcome the not-yet.

The problem for most Western Christians is that we don't seek God as wholeheartedly as we should. We need to learn how to be persistent! One of the reasons why healing doesn't occur as much as we would wish is because we won't persevere in prayer beyond the first minute or two of trying.

In a time of prayer ministry, we might need to be willing to linger a while in the Lord's presence beyond the initial prayers for healing. The *Discern* part of the 5 Step Model is especially relevant. If healing is not happening, we need to go back to God and ask for discernment as to what is going on and where the root of the problem lies.

For some situations there will be multiple times of prayer for healing. We don't know why this is, nor can we tell in advance to whom this will apply. However, we have seen people slowly healed over an extended period of time. For others, it's the '100th' time that prayer ministry takes place where a healing 'suddenly' comes - presumably as a fruit of all those previous times of prayer. It seems that some situations require extended 'soakings' in the presence of God before the release finally comes.

APPLICATION: If you're not seeing breakthrough with someone, does that person need an extended time of prayer ministry, or regular, ongoing prayer?

If they are in a long-term illness, what does it look like for their church community to hold them up in ongoing times of prayer that are both pastorally sensitive and simultaneously full of faith?

3. Unconfessed Personal Sin

In James 5:16 we're instructed, *"Confess your sins to each other and pray for each other so that you may be healed."* James is clearly teaching that one reason for sickness can be unconfessed sin.

This principle is reflected in Psalm 107:17-20, which talks about sickness coming as a result of rebellious sin, and the Lord quickly releasing healing after those people turned back to God in confession and repentance.

In 1 Corinthians 11:27-32, Paul directly challenges each believer to ensure that they come to communion with the right attitude. While this text has been interpreted in multiple ways down the centuries, Paul's main point is that each person should ensure that they don't eat the bread and drink from the cup with arrogance or self-centeredness. Since the power and presence of Jesus is there, to fail to honor the reconciling work of Christ on the cross might lead a believer to *"eat and drink judgment"* on themselves, which in turn might be made manifest through sickness, or even death.

The bigger point from these texts and others is this:

Our unconfessed sin can be a barrier to God's full healing power breaking through.

Sometimes this might be a current sin that we are aware is a problem; other times it might be something from our past that we had ignored, forgotten, or repressed.

When ministering to someone and healing seems stuck, it can be helpful to ask Jesus if there is unconfessed sin that needs to be repented of. We must trust that if this is the case, then He will make that clear - either to the one seeking healing, or to those praying. This is one example of why the prophetic gifts are so foundational, since in this case they might be utilized to ask God for revelation that releases breakthrough.

APPLICATION: If you sense the Spirit saying there's unconfessed sin that's holding back healing, what is a gracious way to invite someone into confession and repentance? Think how you would want to be treated. How can you counteract shame while also leading them to break the hold of the enemy in this area of their life?

4. Lack of Discernment

An incorrect or incomplete diagnosis of the root cause of the sickness might mean people don't know how to pray correctly. For instance, the person being prayed for might be holding on to a destructive lie, or has received a curse that needs to be broken, or perhaps an emotional healing needs to take place before a physical one.

While the Lord is gracious and healing can come even when we totally miss what's really going on, if a situation is stuck, it is important to pause and discern what is the root cause. That way we can pray specifically, as specific prayers tend to receive specific answers.

APPLICATION: When healing hasn't come, consider whether you've taken time to lay aside your initial theory and ask Jesus what is really going on. Have you helped the one seeking healing to do the same?

5. Lack of Submission and Community

James 5:14-15 suggests that the one who is sick should ask the elders to come and pray in faith, and to anoint them with oil. James makes clear that this is a regular and powerful way to experience healing.

This text is not saying that only elders can heal. In fact, the very next verse makes clear that all believers are to minister healing. But James is making an important point about the centrality of being a follower of Jesus within the safety and authority of Christian community.

In the context of the New Testament, elders were always the leaders of the house church (as that was the only way the church gathered from the time of Stephen's martyrdom and throughout the following 3 centuries). You knew them well, and they knew you well, since you were in and out of one another's lives all of the time.

By asking your community leaders to anoint you and pray for healing, you are recognizing their spiritual authority over you, and that Christian life can not be independent of others. The flip-side of this submission is that the elders commit to lead a Christian community that becomes a hot-house environment of expectancy to see prayers answered.

APPLICATION: If you're not seeing healing, what is the response of the church community? Is there an atmosphere of persistent expectancy when prayer occurs? Are the leaders actively pursuing the Lord for the water table of healing to be rising?

Also, if you haven't done so yet, is this the time to call in the ones with more spiritual authority (your elders) to minister healing?

What does it look like for you to fuel a greater level of faith in your Christian family?

6. Lack of Personal Preparation

In Mark 9:28-29, the disciples ask Jesus why they failed to cast out a demon, and He replied, *"This kind can come out only by prayer (and fasting)"*.

Think about what He's saying here. Jesus is not instructing them that they pause the prayer ministry in order to go away and have an extended time of prayer about that specific situation (after all, by definition fasting will take a good chunk of time!). A better way to understand this text is to see that Jesus is talking about personal preparation and spiritual maturing.

Even though we may be deeply committed to Christ, that doesn't mean we are trained or mature enough to meaningfully participate with God in His ministry of extending His Kingdom rule. **While ministering with Jesus is a Spirit-empowered act of grace, Jesus wants us to become meaningful partners with Him. This requires us to mature and grow, including through training and preparation.**

In this passage, prayer and fasting represent an ongoing pursuit of the Lord, not so much for a specific situation (although, of course, that's always a good thing to do), but also more for our general and overall growth and maturation.

As we walk with Jesus in healing the sick, we will experience ebbs and flows of His power and authority. This stops us from falling into presumption (e.g. "Here's the formula for healing backs") and keeps us walking in partnership and dependency upon His specific presence and direction. That doesn't stop us from having higher faith for certain conditions, and we can indeed discern general principles and patterns that often do help and work well, but always in the context of relationship and sensitivity to His voice.

Near the end of his life, Paul urged Timothy to *"be prepared in season and out of season"* (2 Timothy 4:2) to persevere, minister, lead, preach, and evangelize. In other words, we must take responsibility for our preparation - both in our spirit and our skills.

APPLICATION: If you're ministering in a specific context where healing hasn't come, consider if there's someone else the Lord wants to bring into the praying.

More generally, what does it look like to increase our partnership with and dependency on Jesus? How can we deepen our pursuit of Him? What do you need to do to be better prepared in your spirit to be ready to bring healing?

Specifically, what would prayer and fasting preparation look like for you?

7. We Miss The Bigger Picture

In perhaps the most famous of all Bible verses, we are plainly told, *"For God so loved the world that he gave his one and only Son, that whoever believes in him shall not perish but have eternal life."* (John 3:16)

God's highest purpose isn't to heal our bodies, but to save our souls. We see this principle acted out by Jesus in places such as Luke 5:17-26. When the paralyzed man is lowered through the roof in front of Him, His instinctive response was to look beyond the physical and into the spiritual. *"When Jesus saw their faith, he said, 'Friend, your sins are forgiven'."*

How fascinating that when Jesus sees faith, He moves to forgive sin! Of course, His message of the Kingdom also includes healing, and so the man is healed as well. But note that Jesus makes clear that to fully forgive sin is a sign of His authority.

Sometimes when people aren't healed, we need to bear in mind that God might be doing things that we don't see or understand. His highest purpose is not focused primarily on our bodies in this earth, but on our eternal salvation and partnership with Him henceforth. While healing is an expression of this saving activity, it is not the most important thing.

APPLICATION: Are you keeping the bigger picture in front of you? Yes, absolutely pursue healing the sick, but are you guilty of using this as a substitute for evangelism and sharing your faith?

How can you better keep the bigger picture - people committing to Christ as Savior and Lord - front and center?

8. We Don't Know How

We believe that God wants His church to see more success in healing the sick. When this doesn't occur, we can use a list like this one to come back to Jesus and ask for direction in growing in the gifts of healing, and as a result take tangible steps to move forward.

However, we must also recognize that **there will always be an aspect of healing that is a deep mystery to us.** We don't know in advance who will be healed, and we can't control what happens supernaturally. Things don't work out how we would wish. Frankly, at times it is confusing and seems to be hugely unfair for those left unhealed.

This is the tension of living in the now and the not-yet of God's Kingdom. We must not so lean into the 'now' of the Kingdom that we take Jesus for granted and presume to have an infallible plan, nor should we so embrace the 'not-yet' that we become apathetic and fail to do what we have been commanded to do.

Somehow we must learn to hold onto both of these competing perspectives at the same time.

APPLICATION: Can you be at peace with not always having the answer on healing? If not, what is the blockage?

How can you better help those to whom you minister to accept the mystery aspect, while still building up their faith in the goodness of our miracle-working God?

OBEDIENCE OVER DISAPPOINTMENT

We live in a world that is not perfect. There is sickness, injury, and death - none of which are part of God's original plan for creation, nor are they part of our future in the renewed heavens and earth for which we long (Revelation 21:4).

If it's true that *"Jesus Christ is the same yesterday and today and forever"* (Hebrews 13:8), that means He's still the God *"who forgives all your sins and heals all your diseases, and redeems your life from the pit"* (Psalm 103:3). To overcome all of the consequences of the Fall, Jesus has already won the battle over sickness at the cross. *"He took up our pain and bore our suffering... and by his wounds we are healed"* (Isaiah 53:4-5). It is therefore logical to assume that God deeply desires that everyone be healthy and whole.

So when we are faced with outcomes that we did not want, when a mighty battle of prayer and perseverance is fought yet the loved one is stuck in long-term illness or even dies, we have a choice to make. Will we still center ourselves on God's truths and the reality of His coming Kingdom? Will we choose to obey the repeated command of Jesus to go heal the sick, drive out demons, and proclaim the Kingdom?

We must be the ones who refuse to let our disappointments in life redefine our understanding of Scripture. Instead, our understanding of Scripture must redefine our disappointments in life.

We must not bring God's Word down to our level of experience, since that is idolatry, a creating of God in our own image.

Even if we haven't yet seen it, we must persevere and pursue greater faith. As a couple, we have prayed for people who have died, and we have been heartbroken as a result. But those times only serve to cause us to redouble our efforts to fight against the enemy's wickedness made manifest through sickness and death. We choose to pursue greater effectiveness in healing, to be part of a Christian community where faith is growing, to see more and greater healings occur (see John 14:12).

How do we handle it when people aren't healed? We have to be able to live in that tension, pain, and mystery. While we might not have all the answers, we do know, though, that the wrong answer is to give up trying. Instead, we choose to press into the supernatural even more.

If you are the person who is being prayed for in a chronic situation, it is so hard. Our hearts sincerely go out to you. We humbly believe that what you need is for the Christians around you to step out with increasing faith and perseverance, in accordance with the clear instructions of Jesus. This might take a bit of prodding on your part! And while you wait, you need to be surrounded by the love and goodness of God, as lived out through the lives of the believers. *"Carry each other's burdens, and in this way you will fulfill the law of Christ."* (Galatians 6:2)

At the same time, it's also important to remember that if we as believers don't step out to heal the sick, nothing supernatural is likely to happen, and that poor person is stuck in the same situation. Even though none of us will see a 100% success rate, even a 25% success rate (and as a couple we see far higher) is still a huge win for those individuals healed, as well as for the fame of Christ. The more you seek to heal the sick, the more the sick will be healed. Keep persevering!

A FINAL ENCOURAGEMENT

Τhe mandate to heal the sick is meant to be something that is easy and light to carry into life (Matthew 11:30). We're to simply operate out of love and compassion, and calmly command, "Be healed in Jesus' name."

Jesus did not describe healing the sick in opaque mystical terms. The Gospels show Him operating in a very grounded yet faith-filled authority. And His repeated message to His disciples was, "You too can heal the sick!"

Today, Jesus' message to you is, "You too can heal the sick!"

And so our final encouragement to you is to always step into healing with the faith that this is not some giant divine practical joke, and that Jesus will indeed heal the sick through you!

BE SENSITIVE TO THE SPIRIT

We read in Acts 10:38 that Jesus healed *"all who were under the power of the devil."* In the Gospels, everyone who came to Jesus for healing received it. Clearly Jesus had an incredible healing ministry - one that was powerful, effective, and open to anyone who came.

Yet, it is also true to say that not everyone who was sick on the planet at that time was healed. Presumably there were sick people on the streets whom Jesus walked past and yet He did not heal them. This explains, for instance, why the story of blind Bartimaeus stood out to the Gospel writers as worthy of recording, as it was his persistent boldness that caused Jesus to stop and restore his sight. Such a healing would not have been remembered if everyone Jesus walked near was automatically healed.

We need to be obedient to what Jesus actually calls us to do. He did not say 'pray for the sick', but rather, 'heal the sick'. He did not mean the whole planet, but He did mean those whom the Spirit brings to our attention. And so we find ourselves in way over our heads!

Remember that...

1. Success Is Revealing The Father's Love

As we have repeatedly stressed, the actual supernatural healing is not ours to give. But what is in our control is the level of love, compassion, and goodness that the one who is sick experiences. You are Jesus' representative, so always make sure that you operate in accord with His values and character.

The number one goal in any naturally supernatural ministry, including healing the sick, is that the one you serve experiences afresh the Father's beautifully enriching love for them.

2. You Have Standing Orders

As a disciple of Jesus, you're called to be someone who hears and obeys what Jesus is saying. To help you with this, you've been given standing orders to streamline your decision making. These are your default instructions in the absence of any other specific direction.

When it comes to encountering illness and injury, your standing orders are simple: Heal the sick.

Of course, you should do so in a way that ensures the one to whom you minister feels genuinely loved and known by God, even if their ailment has not gone away. But there are so many opportunities waiting for you, if only you'll remember your orders!

3. Jesus Has Given You Authority Over Sickness

You can't heal out of your own strength or holiness (phew!), but you can do this through the authority and power of the Risen Lord Jesus. He graciously delegates this power and authority to you, and instructs you to heal the sick.

When encountering sickness, especially something more serious or long-term, don't be intimidated by it, and don't give it honor. Instead, sickness - like death and sin and all the works of the evil one - must submit to the rule of Jesus.

This means we pray like someone who has the authority to command sickness to leave and healing to come. You don't need to shout or develop a strangely sweaty stare, but rather you can crisply command this thing to be gone.

Remember: you are the one carrying the authority that trumps all the works of the evil one.

4. Persistent Expectancy Is Vital

You are designed to be one who operates with great faith. There is to be a deep, unshakeable active trust that God is indeed good, that He has compassion on us, and He has the means to do something about the sickness that afflicts us. Hence you embrace the opportunity to heal the sick, with a humble expectancy that Jesus will move through you.

Simultaneously, you recognize that you live in a world where the Kingdom is sometimes not-yet present, and thus when sickness remains, you know how to persevere. Whether it is a longer one-off session, or multiple shorter times of ministering to the same person, you chose to persist in the belief that your prayers for the Kingdom to come on earth as in heaven include this need for healing. And while you wait for that 'stuck' healing to come about, you minister healing to as many others as you can along the way.

OBEDIENCE IN THE MYSTERY

If you have ever walked with a friend or family member through a major illness, you'll likely have prayed with them for healing on a number of occasions. The first few times there is energy around the moment, but as it becomes the second, third, fourth or fifth occasion, the faith drops and it feels more like going through the motions. Eventually it becomes less painful all round not to even bring up the option, and very often many will even stop seeking healing for anyone else.

The scenario above can happen to any believer, because it's easy to become worn down by the battle. To keep going in the face of no breakthrough, or as things deteriorate, is so incredibly difficult. The enemy resists with colossal unfairness, and maintaining a dogged persistence in continuing to pursue healing comes at a high emotional cost. It is a perplexing and frustrating mystery. The temptation is thus, everso gradually, to begin to walk by sight rather than by faith.

Yet, through it all, your call is to be obedient to Christ and the clear teaching He gives in Scripture. In the same way that you shouldn't give up sharing your faith when not everyone you know commits to Christ as Savior and Lord, likewise you shouldn't give up on praying for the sick, even though not all are made well.

Let's put it another way: to insist that your mind must understand and approve every single theological issue before you'll pray for healing is arrogance of the highest order.

It is absolutely fine to wrestle with God in prayer about these things, to read widely, and to discuss experiences and thinking with others who are stepping into these things. Yet humility means that you must simultaneously be at peace with the mystery aspect of the supernatural fight in which you're engaged.

Healing is always about bringing God glory. It is an expression of His Kingdom being present now, and satisfies a deep, divinely-placed longing for things to be made right and whole.

This means that you choose to keep-on keeping-on, even when you're forced to navigate through great disappointment and pain. You must never, ever, give up. Choose to obey Jesus, walking in partnership with Him, and trusting for growing numbers of healings to take place.

Make every effort to keep your heart soft and malleable. Cry out to the Lord, fall on your face before Him, and ultimately get yourself to the place where you can surrender your heavy burden before Him who knows and understands it all. There is almost nothing more movingly beautiful than to see those who might have every excuse to give up choose to yield all to Him, and to lift their hearts in simple trust and worship.

Jesus healed because he had great compassion.

People walked away with a greater revelation of God and His love for them personally. Follow His compelling example, and see what Jesus will do through you. Lives will be transformed, the enemy will be driven back, and God's Kingdom will advance. It will be the most wonderful of adventures!

ONE MORE STORY

We had planned to finish with a stellar story of supernatural healing, to wow and inspire you to press on! However, as much fun as they are, sometimes the huge stories can feel intimidating as a model to imitate, especially if you're just starting out with baby steps in healing the sick.

Instead, here's a seemingly everyday account of healing. But notice this: it records someone's first experience of healing the sick. It comes from a young man who used some of the material in this book to learn how to pray for the sick.

> *"Shortly after hearing this teaching, we spent the day with my in-laws. My wife's father mentioned that his back was hurting, and so, taking my life in my hands, I asked him if I could pray for it! I spent*

a few minutes praying for him, and although nothing seemed to change, it was a good thing to do relationally.

But then the next morning he texted to say thank you, and report that his back was feeling much better! This is a new area of discipleship for me, and the prayer wasn't polished or pretty, but thanks to this training, I tried healing the sick and God received the glory!"

PART II

NEXT STEPS

Having focused up to now on giving you teaching and content, this second section gives you lots of ways to put what you are reading and learning into practice.

This will come to you in 8 parts:

1. Activation Exercises - Some simple exercises that you can do solo and with others. These will help train you in, as well as grow your faith for, healing the sick.

2. Common Questions and Objections - We give you 1-2 sentence headline answers to some of the common questions that you may have yourself, and which you'll probably be asked as you train others on this topic.

3. Incorporating Healing Into Church Life - Some initial thoughts on how to see your church develop a healing culture.

4. Group Study Guide - Many of you will be using this book in a group setting, so here you'll find a guide to help stir up your conversation and next steps!

5. Scriptures to Ponder - While healing the sick might feel new and unusual to some, it is deeply anchored in the Bible. Here we simply list some of the key verses that have most shaped our thinking - and invite you to consider them for yourself, as you ask the Lord how He wants you to live them out today.

6. Further Reading - For those who like to read, here are some suggestions to help you along.

7. Prayer to Grow in Healing the Sick - This is a longer, written-out prayer that you might like to use to express your desire to heal the sick more regularly.

8. And Finally - We share how you can connect with us, access more of our free and paid for resources, and continue to step further into a naturally supernatural lifestyle.

ACTIVATION EXERCISES

NEXT STEPS 1

In order to move further into healing the sick, it is important to practice.

Like any gift from God, He does not simply plop healing fully formed into your lap! Instead, it is something into which you must choose to grow. This requires you to dig deeper into Scripture to understand how the gift works, to learn good practice from those who are more experienced, and then recognize how it connects with the unique combination of your personality, story, and context.

Just as an athlete practices a sport and does warm up exercises first, or a musician practices an instrument with routine scales, so spiritually, by doing some fairly simple exercises, we can grow in faith, perseverance, and effectiveness in the realm of healing. This in turn helps us to be ready when the needed moment of healing and Kingdom breakthrough arises.

This requires discipline and experimentation in environments where you can make mistakes and learn, without causing ongoing pastoral damage to others. Walking this journey in community is vital.

Some of that wisdom will come from the wider church and some from your local church. While we as authors can be one of the many voices who enable you to access the wider church's understanding and experience of the healing gifts, we cannot help you be accountable in your specific situation. For that, you need people who can look you in the eye and help you grow, in a spirit of love, encouragement, and accountability.

To move you along in that process, here are some activation exercises. You will see that they come in four contexts: things you can do individually, some you can do in pairs, there are suggestions for children of different ages, and others to be done in a group context. Obviously, these are jumping off points, so please do adapt them as you wish.

Bear in mind that, because these are artificial constructs, they might feel a bit odd to do! The goal is to stretch you, so that you become more comfortable with the process of healing the sick.

INDIVIDUALLY

1. *Grow Your Desire* - Over the next month, take time each day to read the Bible on healing, pray for those you know who are sick, and ask God to release (greater) gifts of healing to you and your church family.

If you aren't sure what to read, use the list in *Next Steps 5 - Scriptures to Ponder.* You could also jump into the first half of any of the Gospels to find many accounts of Jesus healing the sick, while throughout Acts there are multiple testimonies of the early church healing the sick.

2. *Remove Barriers* - Ask God, 'What is holding me back from stepping further into a ministry of healing?'

Sit in His presence and wait for any insights. If there is a hurt, mindset, or false belief that is holding you back, be sure to repent of it and ask Jesus to replace it with the opposite thing. For instance, if you doubt that God really wants to heal through you, allow Him to replace that fear with a resounding confidence that your Heavenly Father is good, compassionate, and full of overflowing love.

WITH OTHER CHRISTIANS

3. *Intentional Practice* - Find a few Christians with whom you can all practice praying for the sick. Coach each other on your technique (the words you say, body language, tone you strike, how you stir up faith, etc). Even if the health issue is something very minor, practice together so at the very least you are comfortable with speaking out healing prayers and listening to God's voice in that moment.

4. *Pray Generously* - Commit to pray for every Christian you encounter who is sick. Start with your small group/ missional community and around your weekend worship service.

FOR FAMILIES

5. *Preschoolers* - From when children are small, model that we always pray for healing. All the little bumps, bruises, and tears of toddler life are great opportunities to do this. Invite children to pray with you when someone is sick, using age appropriate words and length of prayers. Also have them join in laying on hands and showing compassion.

6. *Elementary Age* - As children mature, invite them to take the lead in praying for healing. This could be a family member who is under the weather, or for a guest, or for someone known to you who is ill. Encourage them to learn to pause in the Lord's presence to listen for any specific direction that He might have in that moment.

7. *Teenagers* - For middle and high school ages, seek to encourage them to take the lead in ministering healing outside of the home as well. This could be at school, church, or amongst their friends. One area that we have found works well is when they can join small teams in ministering healing in church services (as part of a prayer ministry team), or on the streets when tied to evangelism. Be aware that most teens will go through a phase when it's much cooler to do this alongside people who aren't Mom or Dad!

IN THE WIDER WORLD

8. *Prayer Walk* - In your neighborhood or place of mission, go on a prayer walk to specifically ask God for His healing power to break out there. If you know of a household struggling with illness, be sure to pray boldly and firmly near there. See if the Holy Spirit reveals anything specific for you to pray into.

9. *Offer Prayer* - When a friend, colleague or neighbor is sick or injured, engage with them about the illness while internally asking the Lord for faith to offer healing prayer! When you speak, do so with brevity, clarity, and boldness. Don't forget: the #1 goal is that they experience the love of God for them.

10. *Healing on the Streets* - Join with some others who are taking prayer for healing into public places. This will teach you about spiritual warfare, evangelism, and not to take too long when praying for healing! Who can you join with? Is there an existing group in your town, or can you gather a few others to start your own outreach?

COMMON QUESTIONS AND OBJECTIONS

NEXT STEPS 2

Think of this section as a quick-fire round, where we try to give 3 sentence headline answers to questions that are often asked on this topic. The aim here is to remind you of the key points, most of which are unpacked in fuller depth in the main chapters.

Is Healing the Sick Really for Every Follower of Jesus?

Yes — everyone who follows Jesus is called by Him to intentionally participate in healing the sick. **It is intended to be normal behavior for all believers.** *"As you go, proclaim this message: 'The kingdom of heaven has come near.' Heal the sick, raise the dead, cleanse those who have leprosy, drive out demons. Freely you have received; freely give."* (Matthew 10:7-8)

What Is Our Top Objective When Praying for the Sick?

Obviously we always want to see people healed but, as supernatural healing is not something that is within our personal gift, it doesn't always take place. However, we can always minister to the one who is sick with kindness, compassion, and faith. Therefore, the #1 goal is that they leave that time having experienced afresh the Father's love for them.

Why Is Healing an Obedience Issue?

"Therefore… pray for each other so that you may be healed. The prayer of a righteous person is powerful and effective." (James 5:16). If we are disciples of Jesus, that means we are committed to hearing and obeying what He tells us to do and be. Since Jesus clearly teaches us to heal the sick, **our response must be one of faith and obedience**, even if that challenges our prior theological assumptions, proves to be more stretching than we anticipated, or we don't see the outcomes for which we long.

Why did Jesus heal?

If Jesus revealed the nature of God the Father, and Jesus regularly healed the sick, then clearly healing is part of the Father's nature and will. God wants to holistically set us free from the effects of sin, evil, and death, and bring us into our eternal destinies, and so healing is a piece of this larger mission. It is a foretaste of the coming Kingdom, and a natural response of our loving and kind Heavenly Father.

What should our motives be?

We should be moved with compassion and love for the one who is sick, and a desire to partner in faith with God in what He wants to do in that situation. We recognize that the root of our spiritual authority is rapid and full obedience to Jesus, and so we always want to represent Him well and bring Him honor. It is when we are willing to humble ourselves and persist in faith that we best position ourselves to be conduits of the Spirit's power.

How do we access Divine Authority and Power for Healing?

We need to understand that through His incarnation, crucifixion, and resurrection, Jesus has already taken back all authority over the works of the evil one, which means sickness is crushed by His healing power. After His ascension, Jesus delegated this authority and power to His disciples. As we submit to Jesus as Lord and King, He becomes the source of all Spirit-led ministry that occurs through us, including supernatural healing.

Whose Faith Is Required for Healing?

Jesus repeatedly tied together faith and healing, although we must be careful not to turn this wisdom into a legalistic formula (since God loves to break our rules!). **This faith can come from the person seeking healing, their friends and relatives, those who pray for healing, or from the church community**. One challenge for a church is to seek ways to corporately raise the water-table of faith for healing, as part of their shared desire to see today more of Jesus' dynamic Kingdom rule in their places of mission.

How Do I Increase My Faith for Healing?

Study the example of Jesus in the Gospels, the Early Church in Acts and the rest of the New Testament (see *Next Steps 5 - Scriptures to Ponder*). Read stories of healings from church history and the global church today (where it is extremely common, especially in the Global South), and spend time with those who minister with humble boldness. And, above all, ask Jesus - He loves that sort of prayer request!

What Is the Simplest Way to Pray for a Healing?

There is no secret, sacred, or magic formula for healing the sick. The irreducible minimum is a declaration similar to, "In the name of Jesus, be healed!". While we must always keep attentive to what the Spirit is saying in each situation, the essence of healing is that simple.

What Is the Five-Step Model?

1. LISTEN — *Identify the Real Issue*

2. DISCERN — *"How should I pray?"*

3. PRAY — *Share the Father's Heart*

4. REVIEW — *"What's going on?"*

5. CLARIFY — *How to Move Forward*

Why Aren't People Always Healed?

There are many reasons, but Jesus seemed especially to focus on two: lack of faith, and lack of persistence. If you sense lack of faith, without in any way bringing condemnation, play your part in stirring up humble expectancy in the ones praying as well as the one who is sick. If the issue is persistence, consider what specific next steps will enable the one seeking healing to receive sustained and ongoing prayer for healing.

What Do You Say When Someone Isn't Healed?

If something good happened, point that out (e.g. their pain has decreased, or they sensed the presence of God), and remind them that sometimes the healing happens over time. Encourage them to be persistent and keep coming to God for prayer. Above all, never cast blame - instead, encourage them that God is still good and full of love for them right here right now.

How Do I Persist When Nothing Seems to Be Happening?

Your choice is to be obedient to Christ's clear command to heal the sick. You can't control that outcome, but you can choose to operate out of a heart of great love and compassion, and to leave that person more aware of the Father's love for them. Keep fueling yourself with Scriptures and stories that remind you of the Father's deep desire to heal the sick, and ask for faith to persevere and see much fruit in this vital ministry.

Communities of faith should be Communities of healing.

INCORPORATING HEALING INTO CHURCH LIFE

NEXT STEPS 3

I n each of our books in the Naturally Supernatural series, we like to offer some suggestions for how to incorporate what you're learning into the life of your church.

1. Ensure That Your Personal Devotion to Christ Is Deepening

Ultimately any healing ministry flows out of our walk as disciple-making disciples of Jesus. This means that in every area of life we commit to being ones who hear and obey what He is saying.

The healing gifts sit the most meaningfully on a person who is overflowing with love for Jesus. While the Lord is gracious and generous to heal even when the one praying has a stinky attitude, we should not take Him for granted. If daily we are pursuing closeness and intimacy with our Heavenly Father, that is the best possible groundwork for healing the sick.

For those in your church who want to heal the sick (which is a good thing to desire), your top advice needs to center around their personal devotion to God.

2. Keep Pursuing This Personally

Ask God to help you have more opportunities to step into healing, as you reflect on your experiences through the lens of Scripture. We'd encourage you to make an active decision to offer prayer to anyone you interact with who needs healing (unless the Lord clearly directs you not to).

As you move in this direction, others will notice and will start to come to you for advice and encouragement on how to heal the sick.

3. Do This in Groups

Make space for prayer for healing in your church's group life. This means that leaders are coached on how to pray for the sick, why it's important to regularly make space for this, and how they can access advice and feedback on how to handle trickier situations.

The goal is that everyone in the group feels able to pray for the sick and expect healing to occur.

4. Build a Prayer Ministry Team

While any believer can pray for healing, you will find it ever so useful to raise up a team of folks who can take it in turns to be 'on duty' at church services, prayer gatherings, and similar events. Hopefully this book will give you plenty of content to form a basis for their training!

5. Public Worship Services

When the church gathers corporately, create spaces where people can respond to the Lord's initiative in their lives. Part of this includes allowing those who are sick to come for prayer for healing from the prayer ministry team.

Sometimes this might form part of your formal service (e.g. during a time of worship, or while receiving communion, or in response to the sermon), and other times it might be an invitation given at the end of the service to stay behind for prayer.

We'd encourage you to experiment with the actual physical location in your building where people can receive prayer. It is surprising how much that can impact people's willingness to make themselves vulnerable in the pursuit of healing.

6. Collect and Tell Stories

There is so much power in testimony, so you want to tell as many stories of healing as possible!

Some of these will come from other places - church history, books you read, overseas missionaries, friends whose churches are further on than yours. However, the best testimonies come from within your own congregation, even if the stories themselves are small or simple. In fact, even an account of someone trying and failing but committing to pick themselves up and try again is gripping and inspiring to others. "If old Bertha can try praying for her neighbor, I guess we should give it a go too!"

Stories can be shared in groups, worship services, church bulletins and emails, on your website, at kids and student groups, at leadership meetings, and in fact anywhere!

7. Don't Give Up

We've found that persevering faith is critical to building a long-term culture of healing. This means you commit to doing this week-in and week-out for as long as it takes.

Some churches will quickly see many healings occur, while other places have to stick at it. We know one person who prayed for over 100 people before he saw the first healing - yet today he is hugely fruitful in this area. In his book *Power Healing*, John Wimber recorded how Jesus told him to start praying for the sick in Sunday worship services, but it took 10 months until he saw the first healing! He recalled how discouraging that was, and how he pleaded with the Lord to be able to stop. Fortunately he didn't, as his ministry and writing became hugely influential.

So each time you pray, go in expecting supernatural healing. If that's what occurs, give God the glory! But if it doesn't happen, refuse to give up. The problem won't be on God's end. For instance, it could be a heart issue in your church culture that needs dealing with, it might be spiritual warfare that must be overcome, or sometimes you will simply have no idea where the logjam is. However, persevering faith is critical - and you will learn so much through that process.

GROUP STUDY GUIDE

NEXT STEPS 4

F or groups wishing to study this book together, here is a study guide for each of the main chapters to help move your conversation in a productive direction.

A few tips:

- We recommend that you read the relevant chapter(s) in advance of your discussion.
- Make sure that you also do the activation exercises (*Next Steps 1 - Activation Exercises*) to ensure that you combine both reasoning and practice in your exploration. It will be fruitful to do some of those group exercises together, as well as processing your responses to the individual exercises.
- You will also want to freely delve into the Bible (*Next Steps 5 - Scriptures to Ponder*) as an anchor for your conversations.
- Don't forget that the goal is not to just have a well-polished theological understanding of healing, but to ensure that you pair that with a Spirit-empowered, persevering, faith-filled community lifestyle of healing the sick.

CHAPTER 1 — OUR HEALING PROBLEM

1. As you read the story of the three women who sought healing, what was your response? More generally, what emotions bubble up in you when supernatural healing is discussed?
2. How do you respond to the assertion that the #1 measure of success when praying for healing should be that the recipient experiences afresh the Father's love for them?
3. For you personally, what is the most compelling reason to pursue healing the sick?
4. What fears or concerns do you have around prayer for healing being used more openly and fully?
5. When have you or a family member ever experienced God's love for you through prayer for healing?

CHAPTER 2 — A BIBLICAL THEOLOGY OF HEALING

1. How do you respond to a story of cancer being miraculously healed?
2. As you look at the examples of Jesus and the Early Church healing the sick, how does that compare to your church? Is their example a fair expectation?
3. In the section on 'What Causes Sickness?', did anything surprise or provoke you?
4. Do you agree that Jesus has given clear standing orders when facing sickness? Why haven't you always obeyed?
5. Do the reflections exercise at the end of the chapter.

Chapter 3 — Principles for Praying

1. Do you believe that any believer can heal the sick? Why?
2. What most blocks you from fully accessing the incredible delegated authority and power that Jesus gives you to heal the sick?
3. Do you tend to use command or request prayers when ministering healing? What would help you pray with greater boldness and authority?
4. Have you ever been in a situation where you sensed the Spirit lead you not to offer prayer for healing? Looking back, was that the right decision?
5. Name three specific practical steps that you will take to increase your faith for healing.

Chapter 4 — A Practical Guide for Healing the Sick

1. When do you think the five-step model will be most useful?
2. How do you stop the five-step model becoming a set of rules, rather than a framework that balances boldness with loving care?
3. When you pray for healing, do you have the faith to pray short but bold prayers? What holds you (or others) back from that approach?
4. What are some helpful things that you've seen done to help those being prayed for to relax and become more open to the Spirit?
5. Do the exercise at the end of the chapter.

Chapter 5 — When the Healing Doesn't Come

1. Why do you think healing still has such an element of mystery about it?
2. Has God ever used sickness to disciple you? As a result, did you want that sickness to continue?
3. What is the best way to balance faith for healing with pastorally sensitive behavior, especially for those with long-term or major illnesses?
4. Review the reasons for understanding why healing doesn't always occur. Which one(s) do you need to most respond to the application questions?
5. How can you (and your Christian community) commit more deeply to obediently pursue healing, even when disappointments hit you hard?

Chapter 6 — A Final Encouragement

1. How would you answer the cautious Christian who says that since we have the full revelation of the Bible, we no longer need the supernatural gifts like healing?
2. What most excites you about the potential of seeing the sick supernaturally healed?
3. On a scale of 1 to 5, how much would you like to exercise the gift of healing the sick? Explain your choice of number.
4. Pray for one another for God to give you greater faith, boldness, and commitment to healing the sick.
5. Minister healing to any amongst you who are sick.

SCRIPTURES TO PONDER

NEXT STEPS 5

W e have tried to root this book strongly in the Bible, since we believe that the naturally supernatural life-style is one that is commanded, commended, and demonstrated there. **As theology is built, it must be shaped by Scripture.**

However, we also recognize that no one is fully correct in their interpretation of Scripture, since we all bring our preferences, blind spots, history, and culture into how we read it. So to help level the playing field a little, here are some texts that we have found particularly helpful in growing in healing the sick.

These are beneficial to read, to study in context, and even to memorize, so that your theology and practice can be as closely allied as possible to the example of Jesus and the early church. Never forget that the Bible is God's universal revelation to all people at all times and in all places - and so it should shape how we pursue a naturally supernatural lifestyle, including healing the sick.

"For I am the Lord, who heals you." (Exodus 15:26)

"Have mercy on me, Lord, for I am faint; heal me, Lord, for my bones are in agony." (Psalm 6:2)

"Lord my God, I called to you for help, and you healed me." (Psalm 30:2)

"Praise the Lord, my soul, and forget not all his benefits - who forgives all your sins and heals all your diseases." (Psalm 103:2-3)

"Some became fools through their rebellious ways and suffered affliction because of their iniquities. They loathed all food and drew near the gates of death. Then they cried to the Lord in their trouble, and he saved them from their distress. He sent out his word and healed them; he rescued them from the grave." (Psalm 107:17-20)

"Do not be wise in your own eyes; fear the Lord and shun evil. This will bring health to your body and nourishment to your bones." (Proverbs 3:7-8)

Speaking of the coming Messiah, *"Then will the eyes of the blind be opened and the ears of the deaf unstopped. Then will the lame leap like a deer, and the mute tongue shout for joy."* (Isaiah 35:5-6)

"But he was pierced for our transgressions, he was crushed for our iniquities; the punishment that brought us peace was on him, and by his wounds we are healed." (Isaiah 53:5)

"I was enraged by their sinful greed; I punished them, and hid my face in anger, yet they kept on in their willful ways. I have seen their ways, but I will heal them; I will guide them and restore comfort to Israel's mourners." (Isaiah 57:17-18)

"But I will restore you to health and heal your wounds,' declares the Lord." (Jeremiah 30:17)

"Nevertheless, I will bring health and healing to it; I will heal my people and will let them enjoy abundant peace and security." (Jeremiah 33:6)

"*Jesus went throughout Galilee, teaching in their synagogues, proclaiming the good news of the kingdom, and healing every disease and sickness among the people.*" (Matthew 4:23)

"*Your kingdom come, your will be done, on earth as it is in heaven.*" (Matthew 6:10)

"*Which of you, if your son asks for bread, will give him a stone? Or if he asks for a fish, will give him a snake? If you, then, though you are evil, know how to give good gifts to your children, how much more will your Father in heaven give good gifts to those who ask him!*" (Matthew 7:9-11)

"*'Lord, I do not deserve to have you come under my roof. But just say the word, and my servant will be healed. For I myself am a man under authority, with soldiers under me. I tell this one, 'Go,' and he goes; and that one, 'Come,' and he comes. I say to my servant, 'Do this,' and he does it.' When Jesus heard this, he was amazed and said to those following him, 'Truly I tell you, I have not found anyone in Israel with such great faith'.*" (Matthew 8:8-10)

"*When evening came, many who were demon-possessed were brought to him, and he drove out the spirits with a word and healed all the sick. This was to fulfill what was spoken through the prophet Isaiah: 'He took up our infirmities and bore our diseases'.*" (Matthew 8:16-17)

"*He touched their eyes and said, 'According to your faith let it be done to you'; and their sight was restored.*" (Matthew 9:29-30)

"*Jesus went through all the towns and villages, teaching in their synagogues, proclaiming the good news of the kingdom and healing every disease and sickness.*" (Matthew 9:35)

"*Jesus called his twelve disciples to him and gave them authority to drive out impure spirits and to heal every disease and sickness.*" (Matthew 10:1)

"As you go, proclaim this message: 'The kingdom of heaven has come near.' Heal the sick, raise the dead, cleanse those who have leprosy, drive out demons. Freely you have received; freely give." (Matthew 10:7-8)

On encountering a man in the synagogue with a shrivelled hand, *"He said to the man, 'Stretch out your hand.' So he stretched it out and it was completely restored, just as sound as the other."* (Matthew 12:13)

"Jesus withdrew from that place. A large crowd followed him, and he healed all who were ill." (Matthew 12:15)

"When Jesus landed and saw a large crowd, he had compassion on them and healed their sick" (Matthew 14:14)

After Jesus walked on the water, *"People brought all their sick to him and begged him to let the sick just touch the edge of his cloak, and all who touched it were healed."* (Matthew 14:35-36)

Before feeding the four thousand, *"Great crowds came to him, bringing the lame, the blind, the crippled, the mute and many others, and laid them at his feet; and he healed them."* (Matthew 15:30)

"Large crowds followed him, and he healed them there." (Matthew 19:2)

After overturning the tables in the temple courts, *"The blind and the lame came to him at the temple, and he healed them."* (Matthew 21:14)

"A man with leprosy came to him and begged him on his knees, 'If you are willing, you can make me clean'. Jesus was indignant. He reached out his hand and touched the man. 'I am willing,' he said. 'Be clean!' Immediately the leprosy left him and he was cleansed." (Mark 1:40-42)

When a paralyzed man was laid before Him, Jesus commanded, *"I tell you, get up, take your mat and go home."* (Mark 2:11)

"[Jesus] could not do any miracles [in Nazareth], except lay his hands on a few sick people and heal them. He was amazed at their lack of faith." (Mark 6:5-6)

"Calling the Twelve to him, he began to send them out two by two and gave them authority over impure spirits." (Mark 6:7)

"They went out and preached that people should repent. They drove out many demons and anointed many sick people with oil and healed them." (Mark 6:12-13)

"Jesus put his fingers into the man's ears. Then he spit and touched the man's tongue. He looked up to heaven and with a deep sigh said to him, 'Ephphatha!' (which means 'Be opened!')." (Mark 7:33-35)

Jesus is called upon to heal a blind man. After initially praying, *"Jesus asked, 'Do you see anything?' [The man] looked up and said, 'I see people; they look like trees walking around.' Once more Jesus put his hands on the man's eyes. Then his eyes were opened, his sight was restored, and he saw everything clearly."* (Mark 8:22-26)

Jesus healed a little boy with the words, *"You deaf and mute spirit, I command you, come out of him and never enter him again."* (Mark 9:25)

"They will place their hands on sick people, and they will get well." (Mark 16:18)

"Then the disciples went out and preached everywhere, and the Lord worked with them and confirmed his word by the signs that accompanied it." (Mark 16:20)

When Peter's mother-in-law had a high fever, *"[Jesus] bent over her and rebuked the fever, and it left her."* (Luke 4:39)

"One day Jesus was teaching, and Pharisees and teachers of the law were sitting there. They had come from every village of Galilee and from Judea and Jerusalem. And the power of the Lord was with Jesus to heal the sick." (Luke 5:17)

Crowds of people had come to hear Jesus and be healed by him, *"And the people all tried to touch him, because power was coming from him and healing them all."* (Luke 6:19)

"At that very time Jesus cured many who had diseases, sicknesses and evil spirits, and gave sight to many who were blind. So he replied to the messengers, 'Go back and report to John what you have seen and heard: The blind receive sight, the lame walk, those who have leprosy are cleansed, the deaf hear, the dead are raised, and the good news is proclaimed to the poor'." (Luke 7:21-22)

"Hearing this, Jesus said to Jairus, 'Don't be afraid; just believe, and she will be healed.'" (Luke 8:50)

"When Jesus had called the Twelve together, he gave them power and authority to drive out all demons and to cure diseases, and he sent them out to proclaim the kingdom of God and to heal the sick." (Luke 9:1-2)

"When you enter a town and are welcomed, eat what is offered to you. Heal the sick who are there and tell them, 'The kingdom of God has come near to you.'" (Luke 10:8-9)

"On a Sabbath Jesus was teaching in one of the synagogues, and a woman was there who had been crippled by a spirit for eighteen years. She was bent over and could not straighten up at all. When Jesus saw her, he called her forward and said to her, 'Woman, you are set free from your infirmity.' Then he put his hands on her, and immediately she straightened up and praised God." (Luke 13:10-13)

Speaking to the Pharisees about Herod, *"He replied, "Go tell that fox, 'I will keep on driving out demons and healing people today and tomorrow, and on the third day I will reach my goal.'"* (Luke 13:32)

At a Pharisee's house, *"There in front of him was a man suffering from abnormal swelling of his body. Jesus asked the Pharisees and experts in the law, 'Is it lawful to heal on the Sabbath or not?' But they remained silent. So taking hold of the man, he healed him and sent him on his way."* (Luke 14:2-4)

Ten lepers approach Jesus and ask Him for healing. *"When he saw them, he said, "Go, show yourselves to the priests." And as they went, they were cleansed."* (Luke 17:11-19)

As Jesus is arrested, the high priest's servant has his ear cut off. *"But Jesus answered, "No more of this!" And he touched the man's ear and healed him."* (Luke 22:52)

A royal official in Cana begged Jesus to heal his son. *"The royal official said, 'Sir, come down before my child dies.' 'Go,' Jesus replied, 'your son will live.' The man took Jesus at his word and departed. While he was still on the way, his servants met him with the news that his boy was living. When he inquired as to the time when his son got better, they said to him, 'Yesterday, at one in the afternoon, the fever left him.' Then the father realized that this was the exact time at which Jesus had said to him, 'Your son will live.' So he and his whole household believed."* (John 4:46-53)

After healing the man at the pool of Bethesda and letting the resulting frenzy die down, Jesus found the man later and said, *"See, you are well again. Stop sinning or something worse may happen to you"* (John 5:1-15). This was not a threat, but rather an explanation of cause and consequence.

"Very truly I tell you, whoever believes in me will do the works I have been doing, and they will do even greater things than these, because I am going to the Father." (John 14:12)

"Then Peter said, 'Silver or gold I do not have, but what I do have I give you. In the name of Jesus Christ of Nazareth, walk.' Taking him

by the right hand, he helped him up, and instantly the man's feet and ankles became strong. He jumped to his feet and began to walk. Then he went with them into the temple courts, walking and jumping, and praising God." (Acts 3:6-8)

"People brought the sick into the streets and laid them on beds and mats so that at least Peter's shadow might fall on some of them as he passed by. Crowds gathered also from the towns around Jerusalem, bringing their sick and those tormented by impure spirits, and all of them were healed." (Acts 5:15-16)

"Philip went down to a city in Samaria and proclaimed the Messiah there. When the crowds heard Philip and saw the signs he performed, they all paid close attention to what he said. For with shrieks, impure spirits came out of many, and many who were paralyzed or lame were healed." (Acts 8:5-7)

Peter met a man in Lydda who had been paralyzed for 8 years. *"'Aeneas,' Peter said to him, 'Jesus Christ heals you. Get up and roll up your mat.' Immediately Aeneas got up."* (Acts 9:32-35)

"God anointed Jesus of Nazareth with the Holy Spirit and power, and… he went around doing good and healing all who were under the power of the devil, because God was with him." (Acts 10:38)

In Iconium, *"So Paul and Barnabas spent considerable time there, speaking boldly for the Lord, who confirmed the message of his grace by enabling them to perform signs and wonders."* (Acts 14:3)

A man lame from birth, *"Listened to Paul as he was speaking. Paul looked directly at him, saw that he had faith to be healed and called out, 'Stand up on your feet!' At that, the man jumped up and began to walk."* (Acts 14:9-10)

"God did extraordinary miracles through Paul, so that even handkerchiefs and aprons that had touched him were taken to the sick, and their illnesses were cured and the evil spirits left them." (Acts 19:11-12)

"[Publius' father] was sick in bed, suffering from fever and dysentery. Paul went in to see him and, after prayer, placed his hands on him and healed him. When this had happened, the rest of the sick on the island came and were cured." (Acts 28:8-10)

"I will not venture to speak of anything except what Christ has accomplished through me in leading the Gentiles to obey God by what I have said and done - by the power of signs and wonders, through the power of the Spirit of God. So from Jerusalem all the way around to Illyricum, I have fully proclaimed the gospel of Christ." (Romans 15:18-19)

"My message and my preaching were not with wise and persuasive words, but with a demonstration of the Spirit's power, so that your faith might not rest on human wisdom, but on God's power." (1 Corinthians 2:4)

"...to another gifts of healing by that one Spirit." (1 Corinthians 12:9)

"[We] walk by faith, not by sight." (2 Corinthians 5:7)

"I persevered in demonstrating among you the marks of a true apostle, including signs, wonders and miracles." (2 Corinthians 12:12)

Paul wrote about his own experience of sickness. *"As you know, it was because of an illness that I first preached the gospel to you, and even though my illness was a trial to you, you did not treat me with contempt or scorn."* (Galatians 4:13-14)

"For we know, brothers and sisters loved by God, that he has chosen you, because our gospel came to you not simply with words but also with power, with the Holy Spirit and deep conviction." (1 Thessalonians 1:4-5)

"May your whole spirit, soul and body be kept blameless at the coming of our Lord Jesus Christ." (1 Thessalonians 5:23)

"Is anyone among you sick? Let them call the elders of the church to pray over them and anoint them with oil in the name of the Lord.

And the prayer offered in faith will make the sick person well; the Lord will raise them up. If they have sinned, they will be forgiven. Therefore confess your sins to each other and pray for each other so that you may be healed. The prayer of a righteous person is powerful and effective." (James 5:14-16)

"This is the confidence we have in approaching God: that if we ask anything according to his will, he hears us. And if we know that he hears us—whatever we ask—we know that we have what we asked of him." (1 John 5:14-15)

"Dear friend, I pray that you may enjoy good health and that all may go well with you, even as your soul is getting along well." (3 John 1:2)

"'[Jesus] himself bore our sins' in his body on the cross, so that we might die to sins and live for righteousness; 'by his wounds you have been healed'." (1 Peter 2:24)

In heaven our risen Lord Jesus will, *"'Wipe every tear from their eyes. There will be no more death' or mourning or crying or pain, for the old order of things has passed away."* (Revelation 21:4)

FURTHER READING

NEXT STEPS 6

Over the years we have read many books on growing in the naturally supernatural. Our aim here is to give you a list that is not overwhelming, yet does have a good mixture of books to help you develop in this area. We have not included books that are out of print or hard to source in the United States.

As you know, this book is part of a series, and initially we envisioned a unique reading list for each individual book. However, many authors cover a variety of topics in their work, and so it seemed simpler to create one master list for the entire series (albeit with occasional one-off additions for certain books in the series).

These are in author alphabetical order, and with a comment or two to introduce each book to you.

————

These are in author alphabetical order, with a comment or two to introduce each book to you.

- **Ruth Haley Barton,** *Invitation to Solitude and Silence* **(2004)** - One woman's journey into the necessity of regular times of withdrawing to be with Jesus. Each chapter ends with a practice or exercise to try out, which gives the book a healthy focus on application.

- **Mike Bickle,** *Growing in the Prophetic* **(1995)** — From someone who pastored a church with a prominent and yet at times chaotic prophetic ministry, those lessons learned create an insightful and practical resource.

- **Christoph Blumhardt,** *The Gospel of God's Reign* **(2014)** — 19th Century German theologian, also a prominent evangelist, faith healer, and politician, his writings focus on bringing God's Kingdom around us by all means possible. Stimulating, even if you don't agree with everything he says!

- **Shawn Bolz,** *God Secrets* **(2017)** — Engaging teaching on developing the gift of words of knowledge, including content on what to do when you get it wrong. Shawn Bolz has a very public track record of operating in this gift, and does a good job demystifying its usage.

- **Shawn Bolz,** *Translating God* **(2015)** — A down-to-earth yet inspiring read that focuses on training people in how to step into the prophetic gifts, with some great stories, a high level of transparency from the author, and teaching that is so practical.

- **Michael Cassidy,** *Bursting the Wineskins* **(1983)** — Written from an African perspective by the man who is the Honorary Chair of the Lausanne Movement, Michael uses his story as a framework to teach Biblically about entering into life in the Spirit.

- **Dave Clayton,** *Revival Starts Here: A Short Conversation on Prayer, Fasting and Revival for Beginners Like Me* **(2018)** — Practical and non-guilt inducing challenge to step more into fasting and prayer, with lots of application ideas. A great short read!

- **Jack Deere,** *Surprised by the Power of the Spirit* **(1993)** — The inspiring story of how a cessationist seminary professor had his life and ministry turned upside down as he experienced the power and presence of the Holy Spirit.

- **Don Dickerman,** *When Pigs Move In* **(2009)** - A very practical book on deliverance, that contains a host of testimonies, followed by lots of details on the nuts and bolts of this vital ministry, all in an easy-to-read style.

- **James Dunn,** *Jesus and the Spirit* **(1975)** - One of the defining scholarly works on the work of the Spirit, this is very readable and engaging. It does an excellent job of unpacking the Spirit experiences of Jesus and the Early Church. (Dunn was also Alex's NT professor at university!)

- **Gordon Fee,** *God's Empowering Presence* **(1994)** - Fee was one of the first Pentecostals to earn a PhD in Biblical studies, and he combines the two streams in this book by literally exegeting every reference to the Spirit in Paul's writings - but the result is anything but a dry academic read.

- **Michael Green,** *Evangelism in the Early Church* **(1970)** - A fascinating read, with all sorts of nuggets that reveal how the Early Church was so effective in reaching others with Gospel. This includes being a Spirit-filled people, and the use of the gifts is well recorded.

- **Bill Johnson,** *God is Good* **(2016)** - An uplifting read about trusting in the goodness of God as revealed in Scripture, so that we in turn can reveal His goodness in the power of the Spirit to a broken world.

- **Bill Johnson & Randy Clark,** *The Essential Guide to Healing* **(2011)** - An excellent book on how to heal the sick, with practical teaching, stirring stories, and sensible wisdom for developing this ministry in a church context.

- **Charles Kraft,** *Defeating Dark Angels* **(2016)** - Clear, Biblically grounded teaching on how demonic oppression takes place, and how to minister deliverance. Kraft has many books, and we found this one to be very helpful and practical.

- **George Eldon Ladd,** *The Gospel of the Kingdom* **(1959)** - A hugely influential book, which explores the mystery that the Kingdom of God is both now and simultaneously not yet, and how the Spirit empowers us to go out on mission to extend God's kingly rule.

- **George Eldon Ladd,** *The Presence of the Future* **(1974)** - The big idea in this book is that the breaking in of the present, dynamic rule of God is the central concept behind Jesus' message and mission - and thus should be for ours.

- **Francis MacNutt,** *Healing* **(1974 - although look for the updated version)** - This was the first modern-era book on healing that has (deservedly) been widely read, and you can see its influence still today. Very practical and packed full of nuggets of wisdom.

- **Charles Price,** *The Real Faith* **(1930s)** - After experiencing baptism in the Spirit in the 1920s, his ministry was transformed and saw incredible healings. This short book is the best reflection on the nature of faith and the naturally supernatural life that we have found.

- **Derek Prince,** *They Shall Expel Demons* **(1998)** - Prince wrote numerous books on deliverance, and this is an excellent primer into this area. Sensible, Biblical, faith-filled, it contains wise teaching and helpful stories gained from many years of experience.

- **David Pytches,** *Come Holy Spirit* **(1994)** - Written more like a logical, list-driven, logistical handbook, this was such a help in our early years of ministering in the power of the Spirit, as it gives you all the major points in a systematic way.

- **Jon Ruthven,** *On the Cessation of the Charismata: The Protestant Polemic on Postbiblical Miracles* **(1993)** - A brilliant deconstruction of cessationism, written from a scholarly Biblical perspective, full of close exegesis of texts and clear arguments.

- **Jordan Seng,** *Miracle Work* **(2012)** - A great overview of ministering in the spiritual gifts, with each teaching chapter followed by a short story chapter, which makes it all feel very grounded and attainable.

- **Sam Storms,** *Practicing the Power* **(2017)** - This book does a great job of showing how stepping into the spiritual gifts is deeply rooted in Scripture. It's especially useful for those who come from a Reformed Calvinist perspective, which is the author's background.

- **Jerry Trousdale,** *Miraculous Movements: How Hundreds of Thousands of Muslims Are Falling in Love with Jesus* **(2012)** - We both love this book! It is packed full of inspirational stories of how the church is growing globally AND gives tools that we can use here in the West.

- **Jerry Trousdale & Glenn Sunshine,** *The Kingdom Unleashed* **(2018)** - Revealing insights on how the church in the Global South is growing rapidly through Disciple-Making Movements. Lots of takeaways on living the principles of Acts today.

- **Kris Vallotton,** *Basic Training for the Prophetic Ministry* **(2014)** - A down-to-earth, clear and helpful training in the prophetic gifts, which feels as if a very fatherly member of your church is steering you into greater maturity!

- **Mark & Pam Virkler,** *How to Hear God's Voice* **(2005)**
  - Designed more as a workbook (with lots of note taking space), it contains Biblical teaching and helpful exercises, and in particular a focus on encountering God through waiting on Him.

- **David & Paul Watson,** *Contagious Disciple Making* **(2014)** - Learning from Disciple-Making Movements across the globe, the stories here are fabulous, and there is much content on how to make disciples in a naturally supernatural way.

- **Dallas Willard,** *The Divine Conspiracy* **(1998)** - An engaging and thoughtful study about the nature of the Gospel of the Kingdom that Jesus preaches - which is not a set of rules to follow, but a declaration of God's active rule and His invitation for us to enter in and partner with Him.

- **John Wimber & Kevin Springer,** *Power Evangelism* **(1985)** - Explains Jesus' theology of the Kingdom, and then moves to show how this transforms our evangelism, with lots of practical stories and ideas for taking your next step.

- **John Wimber & Kevin Springer,** *Power Healing* **(1987)** - A classic text that has influenced many leaders. It creates a Biblical theology of healing, based on how Jesus operated, and then applies those principles with wisdom and experience.

- **Brother Yun,** *The Heavenly Man* **(2002)** - An inspirational first-hand account from Chinese house church leader Brother Yun, who has led a huge movement in the power of the Spirit in the face of severe persecution. Some amazing stories!

13

PRAYER TO GROW IN HEALING THE SICK

NEXT STEPS 7

D ear Heavenly Father,

Thank You for Your invitation to join You in healing the sick! I am amazed and humbled that you would use me to bring healing, hope, and love into the lives of others.

Fill me with a heart of compassion and mercy for those who battle illness. May I learn to respond not only with kind deeds, but also with healing power. Open my eyes to see what it is You want to do in every situation, so that I can walk in step with Your Spirit in unlocking permanent healing and life.

The Bible repeatedly reminds me that, *"By his wounds you have been healed"*. Thank You Jesus for the incredible victory You won on the cross! You have all the authority and power, and nothing is outside of Your reach. It is incredible that You share Your authority with me, so that I might join You in advancing Your kingly rule in the lives of those around me. Teach me to be a faithful steward who is increasingly effective in healing the sick You allow to come across my path.

Your word says, *"When you enter a town and are welcomed, eat what is offered to you. Heal the sick who are there and tell them, 'The kingdom of God has come near to you.'"* Grant me boldness and faith to live like this! May I carry Your healing not just to believers, but also to my lost friends, neighbors, colleagues, and family. May I be used to significantly advance the Kingdom as I obey Your command to heal the sick.

When I encounter the mystery of healings not occurring when or how I would wish, please keep my heart tender and full of faith. Where the call is to walk with someone through a long-term condition, may I be a source of life, hope, and faith for them and those who love them. May I respond with an unshakeable persistence of faith, so that even seemingly intractable illness is broken and driven out of people's lives, however long that takes.

I recognize that it is easy to become distracted by fear, doubt, and the cares of this world, so I ask for the grace to keep my heart and mind locked onto You. Grant me the wisdom not to expect an explanation for every scenario. When the enemy tries to draw me away, may I choose to be still, and again know that You are my God. Keep me humble in the moments where supernatural healing is clearly taking place, so that the glory and my dependency remain fixed upon You.

Thank You for the body of believers around me. Please provide women and men from my church community who are also on this journey of healing, so that together we can raise the water table of faith. Please give me people who can help me grow in this gift through accountability, encouragement, and a shared pursuit of the Lord Jesus Christ.

As I look to step further into healing gifts, may I operate out of a heart of love. Please help me to lay down my agendas, and instead serve in ways that honor You and demonstrably value the people with whom I pray. May I minister out of a humble, submitted heart of compassion for others, and bring healing to many in ways that truly reflect Your great love, kindness, and goodness.

In Jesus' name I pray,

Amen.

The heart of a healer is grounded in the heart of the Father.

14

CONNECT WITH US AND ACCESS MORE!

NEXT STEPS 8

F irstly, THANK YOU for taking the time to read this book! Our hope and prayer is that you now feel greater confidence in healing the sick, and that you can step further into a naturally supernatural lifestyle.

However, the key is not to stop now! If you would like to see your investment of time, energy, and prayer produce a far greater return, then you will need to keep practicing and learning. And we are here to help you on your journey...

ACCESS MORE RESOURCES

As a couple our call from God is to equip the wider church with practical tools like this book. To help facilitate this, we lead the team at Dandelion Resourcing.

At Dandelion, we focus on three core areas:

- Disciple-making
- Living on mission
- Being naturally supernatural

We find that when all three of those circles of life overlap, a Kingdom culture is formed that leads to dynamic life, impact, and growth.

To help ground this into practice, and with passionate believers like you in mind, we have developed a range of resources:

HEALING THE SICK - THE VIDEO COURSE

Ideal for a Church or Group

- Would you love it if you and your church family could more consistently heal the sick?
- Are you held back by a lack of teaching on this topic that is Biblical, practical, and yet faith-filled?
- Do you struggle to know what to do in the more complex situations, or what to say when the healing doesn't come?

If you answer 'Yes' to any of these questions, then the *Healing the Sick Course* will be ideal for you and your church family!

Running over 4 weeks, with sessions designed to last between 60 and 90 minutes, each time you will experience 2 video teaching blocks from Alex and Hannah, totaling around 40-45 minutes, activation exercises, guided discussions, and home-work setting and review.

This practical course unpacks how Jesus heals through His people today in naturally supernatural ways. You will experience clear explanations, Biblical insight, inspiring stories, and tangible next steps that help everyone grow in the healing gifts that the Scriptures clearly command us to pursue.

To find out more, visit dandelionresourcing.com/healing-the-sick-course.

NATURALLY SUPERNATURAL COACHING COHORT

Ideal for Leaders/Church Staff

The best investment you can make in yourself is to join one of our *Naturally Supernatural Coaching Cohorts*, where over 12 months you will exponentially grow in your competency and confidence in living as a naturally supernatural Kingdom leader!

Meeting as part of a small group twice a month on Zoom, we will pour into you a custom mix of theological, philosophical, and practical training. The focus throughout will be on personal application, so that you translate principles into action, build your own back-catalog of stories and experiences, and become a leader in naturally supernatural living.

Topics covered include:

- Hearing the Voice of God
- Healing the Sick
- Building a Naturally Supernatural Theology
- Being Filled with the Holy Spirit
- Operating in Words of Knowledge and Wisdom
- Fighting to Win at Spiritual Warfare
- Discerning and Driving Out Demons
- Building a Culture of Healing
- Living a Miraculous Life of Faith
- Becoming a Naturally Supernatural Missionary
- Building a Prophetic Culture
- Raising Naturally Supernatural Kids, and more!

To find out more, go to dandelionresourcing.com/naturally-supernatural-cohort.

CUSTOMIZED COACHING

The best leaders always benefit from individual coaching, both to help them develop further and also to reach specific goals. We have tons of experience in equipping leaders like you to gain breakthrough in disciple-making, living on mission, or naturally supernatural living. Contact us to discuss what would work best for you in your current situation!

CHURCH CONSULTATIONS

We also do a very limited amount of church consulting, where you and your team can benefit from the breadth of our experience and equipping. We will work with you to craft a boutique process, so that you see sustained culture shifts and growth.

GUEST PREACHING OR TEACHING

If you would like one or both of us to preach or teach at your church, special event, or conference, we would be honored to hear from you! It doesn't need to be on naturally supernatural topics - we have lived, written, and spoken extensively on discipleship, missional living, marriage, and parenting. Whether you're looking for something recorded on video or done live in person, please do reach out to us with your needs.

FREE VIDEO TRAINING

Every couple of weeks we release a short, highly applicable video training. We focus on our three core areas - disciple-making, living on mission, and being naturally supernatural. These are always free and we encourage folks to use them to train and equip others locally. They're also a great opportunity for you to hone your English accents as you listen to us chat!

If you drop your name and email in the sign-up box on dandelionresourcing.com, we'll shoot you an email each time a new video is released (and you can one-click unsubscribe at any time).

MORE WRITING

This book is part of *The Naturally Supernatural Series*, each of which combines Biblical theology, field-tested step-by-step teaching, and repeatable personal practices.

Topics we will cover in the series include the prophetic gifts, healing, deliverance, building a Kingdom theology, being filled with the Holy Spirit, living a miraculous life, and becoming a naturally supernatural missionary.

As a gift from us to you, at the end of this book you will find a free sample chapter of one of our other books, to build on what you have already been learning and implementing!

Share Your Feedback

If you have any questions or suggestions about the content of this book, please feel free to be in touch with either of us. If you go to dandelionresourcing.com, there is a 'Contact Us' form on the site.

We're also on social media at the following spots:

Instagram: instagram.com/alexabsalom

Twitter: twitter.com/alexabsalom

Facebook: facebook.com/dandelionresourcing

Could You Help Us Out?

If you have enjoyed this book, would you mind doing us a quick favor?

The best way for others to find out about a resource like this is through personal recommendation (think about what makes you investigate a new book!). With that mind, we would be so honored *if you would take a moment to share a quick review with others.*

The #1 place is on Amazon - pop in your stars and write a comment, and that will help us enormously. (As you no doubt know, the more reviews, the more their algorithms will highlight this book to others.)

In addition, please do share about this book (post a photo of you holding the cover!) on social media - feel free to tag us in the shot!

Thank you so much in advance.

Again, thank you for reading - we're praying that you have so much Jesus-honoring fruitfulness as you step further into a naturally supernatural lifestyle.

With love and blessings,

Alex and Hannah

ACKNOWLEDGMENTS

Over the years so many people have helped us grow into a naturally supernatural lifestyle. Some have been leaders, others friends, and yet others would never know our names or faces, but have shaped us through their teachings, books, podcasts, and lives.

However, a number of specific people have intentionally helped us with this book - and we are thankful for each one of them:

- Caity Shinnick designed the cover and did all the internal layout work, which turned our chicken scratch into something lovely to look at.

- Marshall Benbow, Mark Burgess, Amy Honeycutt, Rachel Judy, Rich Rollins, and Lee Simmons gave us very helpful feedback and edits as they read through draft copies of the text - thank you all.

- Our Naturally Supernatural Coaching Cohorts, where we together worked through a lot of this material in its early forms.

- Joanne and the baristas at our local Starbucks, where Alex did a lot of his writing. Thank you for creating such a friendly, kind, and creative atmosphere!

Finally, to our boys, Isaac, Samuel, and Joel (yes, Isaac can go first this time!) - thank you for giving us space to create and write, for being such encouraging voices to us, and for living out the principles of this book. We love you and are so proud of you!

BONUS: FREE TASTER FROM 'HEARING THE VOICE OF GOD'

Now that you've finished Hearing the Sick, we invite you to check out this sample from the sister book, Hearing the Voice of God - also available at Amazon or wherever you purchased this book.

There is so much confusion about the prophetic gifts, so let's begin by creating a clear definition. Further on in this chapter we'll share some language that others use, but, for now, here is what we mean:

> *Prophecy is the loving supernatural ability to know and appropriately speak the mind of God on a specific subject at a specific time by the prompting and inspiration of the Holy Spirit.*

A few things to note:

- Prophecy is a **supernatural** ability - it comes from God and is to glorify God, through both the process (how we prophesy) and the outcome (what we prophesy).

- It is rooted and grounded in **love. To share accurate prophetic words without love is to miss the point entirely.**

- When we talk about knowing the **mind of God**, this is not a universal statement - no human could possibly know more than a droplet of what God knows! Yet **we can have the mind of Christ - God's insight, wisdom, and understanding - on a specific matter**. Paul wrote about this in 1 Corinthians 2. His focus in that chapter is on ensuring that the church's faith is not built upon human wisdom, but divine power (v.5). He states clearly that part of our task is to declare and demonstrate God's wisdom - which quite clearly we can't do out of our own cleverness, skills or strength.

- The way we are to access that divine wisdom is **through the Holy Spirit** revealing these things to us. As Paul wrote in v.11, *"For who knows a person's thoughts except their own spirit within them? In the same way no one knows the thoughts of God except the Spirit of God."* Christian prophecy is something that is only possible when it is prompted and inspired by the Holy Spirit.

- This expression of the wisdom of God in prophecy **is specific** rather than universal. It is for a specific person or community in a specific place at a specific point in time. This is one of the ways that it is different from, and inferior to, Scripture. The Bible is God's wisdom revealed by the Spirit to all people in all places at all times. So to believe in the use of the prophetic gifts in no way undermines our high value on the Bible. In fact, done right, it only builds our value for Scripture since, as we'll discover, the Bible is not only the primary way God speaks to us, but is also our first benchmark for evaluating a prophecy.

KEY TEXTS

1 Corinthians 14:1 and 3 are critical texts for creating a framework for our pursuit of prophecy.

"Follow the way of love and eagerly desire gifts of the Spirit, especially prophecy... the one who prophesies speaks to people for their strengthening, encouraging and comfort."

Out of those verses, and allied with some nearby passages, we can draw some more foundational insights about the nature of prophecy.

1. PROPHECY IS ALL ABOUT LOVE

In 1 Corinthians 13:2, Paul writes: *"If I have the gift of prophecy and can fathom all mysteries and all knowledge... but do not have love, I am nothing."*

The point of prophecy is that it is all about love - meaning the thing that is patient, kind, not envious, not proud, and so on!

The enemy tries to bring guilt, shame, and condemnation, but we should bring love.

This doesn't mean that prophecy can't challenge or provoke us to repentance and change (it often does!), but it must always come from a heart of love in us (if we are sharing), and reveal the Father's heart of love for that person or community (so the receiver is aware of that love).

Galatians 5:6 tells us that *"The only thing that counts is faith expressing itself through love."* Often people feel anxiety about prophecy, particularly where it seems strange or spooky. We can help calm that by a tone that is kind and grace-filled, ensuring that what we share reveals God's loving Father heart.

Interestingly, Paul tells us (1 Corinthians 14:1) that we follow the way of love when we *"eagerly desire"* the gifts of the Holy Spirit. The English term 'gifts of the Holy Spirit' in the original Greek is actually something much simpler - literally, 'the spiritual'. English translations add the word 'gifts' to help unpack the meaning. But in essence, Paul tells us to 'eagerly desire the spiritual stuff'!

According to Paul, we perhaps most follow the way of love when we prophesy. Now that is a ministry-changing insight!

Shawn Bolz comments on why Paul so highlights prophecy out of all the gifts: "It can be one of the clearest validations of the Father's great love, which Jesus paid such a high price for. When people hear the thoughts and emotions of God toward them, they believe in his love for them."

2. PROPHECY CONNECTS US WITH GOD

The point of prophecy is to connect us to the Father. Paul writes,*"I keep asking that the God of our Lord Jesus Christ, the glorious Father, may give you the Spirit of wisdom and revelation, so that you may know him better"* (Ephesians 1:17).

It is easy to miss what is going on here. Paul is not asking for us to grow in gifts of wisdom and revelation (good though that is). Instead, the prayer is for us to experience more of God's Spirit of revelation and wisdom, so that we can know the Father better.

Insights into the character, heart, and mind of God can only come by the anointing of the Holy Spirit. And the goal of revelation from God is that we, or others with whom we share revelation, may know the Father more deeply.

Bear in mind that in the biblical understanding, knowledge of God is never in the abstract - it is always tied to obedience and intimacy. If you don't do what He says, you don't know Him, however much information about Him might be in your head.

Although prophecy brings some beautiful and profound insights into our lives, ultimately that is not its purpose. The point is to know and love God, as the Holy Spirit reveals more about the Father to us and through us, to His great glory.

3. PROPHECY IS OPEN TO ALL

Prophecy is something that anyone who is committed to Christ can do. While we won't all have the same level of clarity, regularity, or insight, nevertheless the prophetic gifts (by which we mean prophecy, words of knowledge, words of wisdom, distinguishing between spirits) are open to ALL of us!

Think of it like this: it would be so cruel and deceptive of Paul to urge us all to *"eagerly desire gifts of the Spirit, especially prophecy"* if lots of us were automatically excluded from accessing them! Clearly he isn't doing that - and thus we conclude from Scripture (as well as from experience) that **anyone who loves the Lord Jesus can learn to prophesy**.

Paul's instruction to eagerly desire to prophesy flows directly out of Chapter 13's masterpiece on love. If we view the love passage as being mandatory for all followers of Jesus, by what strange twist of exposition does the start of Chapter 14 on prophecy abruptly stop being for everyone? Paul deeply ties the pursuit of love to the pursuit of the gifts of the Spirit.

Eagerly desiring the gifts is an invitation to everyone who follows Jesus.

4. PROPHECY IS A CHOICE

To eagerly desire means that we CHOOSE to learn, practice, and intentionally grow in our skills in this area. It is a partnership between us and the Spirit.

Candidly, we have met many people who simply don't want to access these gifts! That might seem extraordinary to you, but it is far more common in some church circles than you imagine.

* PAUSE and consider: on a scale of 1 to 5, how eagerly do you desire that you might prophesy?

5. PROPHECY IS TO STRENGTHEN, ENCOURAGE AND COMFORT

Prophecy is for *"strengthening, encouraging and comfort."* This is our non-negotiable ground rule for anyone starting out and becoming established in the prophetic gifts.

And it is such an invitation! After all, wouldn't you like to be more strengthening, encouraging and comforting to others? Wouldn't you like it, when you're helping, advising, and leading others, for it to be less about your own wise words and instead for God to speak through you right into the heart of the situation? Wouldn't you like a reduction of the pressure on you to come up with something 'good' to say? This is a gift to both the hearer and to the sharer!

Kris Vallotton says, "True prophetic ministry looks for the gold in the midst of the dirt of people's lives." In other words, there are no prizes for spotting the junk and the dysfunction. Instead, if we start with how the Father sees them, who He wants them to become, isn't that spirit of encouragement far more likely to bring out the sort of sincere repentance and profound life change that is characteristic of encountering Jesus?

6. PROPHECY CAN BE HINDSIGHT, INSIGHT OR FORESIGHT

Prophecy itself works in three time frames. It can be **hindsight about what we already know, insight into the present, or foresight into the future**.

- **Hindsight** is generally either to gain our attention (often this is called a Word of Knowledge), or to help us look back with better understanding, discernment, and wisdom.

- **Insight** is to help us make sense of what is going on around us, and in particular what the Father is saying and doing, so that we can follow His lead more closely.

- **Foresight** tells us what could be if we partner with the Lord, and thus is often focused on our preparation. Sometimes it tells us about things that God has in mind to bring about.

7. PROPHECY HELPS FORM DISCIPLE-MAKING DISCIPLES

Prophecy is closely tied to being a disciple of Jesus. If you recall, Jesus is at pains to point out that being a disciple means that we hear and obey what He is saying (e.g. Luke 6:46-49). In the New Testament…

Discipleship is always obedience-based.

However, in order to obey we must know what Jesus is saying specifically about our personal context. While some situations are fully covered by God's universal principles (e.g. 'Can I pocket this pile of cash that is sitting on my colleagues' desk?'), that is not always the case.

What about those times when we're wanting to follow biblical principles, but there are several valid options for the situation at hand? For example, do you follow the advice of Psalm 27:14 and *"Wait patiently for the Lord…"*, or do you instead go with Philippians 3:12 and *"…press on to take hold of that for which Christ Jesus took hold of me?"*

We each have times where we are trying to discern between more nuanced options, all of which might be morally good, but yet will lead us down very divergent paths (e.g. Should I take this new job? What school should our child go to? Is it wise to take out a loan to do this additional training?)

These are the bread and butter issues of applied discipleship. And it is in these circumstances that we want to hear if Jesus is saying something specific to us. Sometimes this is by us hearing God's voice for ourselves, and sometimes it is through prophetic words from others. And, of course, sometimes we will be that external prophetic voice for those around us.

There are plenty of times where Jesus is perfectly happy for us to demonstrate maturity and make the choice ourselves. After all, He is seeking to build a family full of sons and daughters, not a factory full of slaves. However, we all recognize that there are also times where we definitely need His leading, or where He wants to more overtly steer us.

Yet how can we follow and learn from Jesus if we don't know how to hear His voice? This is why this is such a vital skill!

8. PROPHECY COMES AS REVELATION, INTERPRETATION AND APPLICATION

We have found that generally **a prophetic word has three parts — revelation, interpretation, and application**.

- **Revelation** is about what Jesus is actually saying to us, and the mechanics of how He does that.

- **Interpretation** is about weighing what has been sensed, and discerning if and how much of it is from Jesus, and what it might mean to us today.

- **Application** is about deciding what I am going to do in response. Will I respond in faith and obedience? What is the timeline? What are my next steps?

Much of the rest of this book unpacks those three headings - each will have its own chapter.

Our book *Hearing the Voice of God* is packed full of Biblical teaching, thoughtful theology, inspirational stories, and practical coaching on the mechanics and details of hearing God's voice and stepping into the prophetic gifts. Like this book, it also includes Next Steps sections to help you personally apply and live out what you will be learning!

ABOUT THE AUTHORS

Alex and Hannah Absalom lead Dandelion Resourcing, which empowers Christians to go and form disciple-making disciples of Jesus in naturally supernatural ways. Originally from England, they have been in church leadership since 1994, live in Long Beach CA, and with their 3 young adult sons are missionaries to the USA.

facebook.com/dandelionresourcing

twitter.com/alexabsalom

instagram.com/alexabsalom

HEALING THE SICK: THE VIDEO COURSE

THE PERFECT WAY TO EQUIP YOUR GROUP OR CHURCH TO
HEAL THE SICK!

Video Teaching, Discussion, Homework, and More

4 Weeks • 60-90 Minute Sessions

Option to have either Alex or Hannah do a live Q&A!

dandelionresourcing.com/healing-the-sick-course

———

Made in the USA
Monee, IL
10 April 2021

65322780R00129